Plant Sensing
and Communication

INTERSPECIFIC INTERACTIONS

A Series Edited by John N. Thompson

Plant Sensing
& Communication

RICHARD KARBAN

The University of Chicago Press | Chicago and London

Richard Karban is professor of entomology and a member of the Center for Population Biology at the University of California, Davis. He is coauthor of *Induced Responses to Herbivory*, also published by the University of Chicago Press, and *How to Do Ecology: A Concise Handbook*.

The University of Chicago Press, Chicago 60637
The University of Chicago Press, Ltd., London
© 2015 by The University of Chicago
All rights reserved. Published 2015.
Printed in the United States of America

24 23 22 21 20 19 18 17 16 15   1 2 3 4 5

ISBN-13: 978-0-226-26467-7 (cloth)
ISBN-13: 978-0-226-26470-7 (paper)
ISBN-13: 978-0-226-26484-4 (e-book)
DOI: 10.7208/chicago/9780226264844.001.0001

Library of Congress Cataloging-in-Publication Data
Karban, Richard, author.
    Plant sensing and communication / Richard Karban.
        pages cm — (Interspecific interactions)
    ISBN 978-0-226-26467-7 (cloth : alk. paper) — ISBN 978-0-226-26470-7 (pbk. : alk. paper) — ISBN 978-0-226-26484-4 (e-book) 1. Plant physiology. 2. Plant ecology. 3. Plant behavior. 4. Plant communication. I. Title. II. Series: Interspecific interactions.
    QK771.K37 2015
    581.4—dc23

2014040550

♾ This paper meets the requirements of ANSI/NISO Z39.48-1992 (Permanence of Paper).

Contents

Acknowledgments

I wrote this book to help me place my own work on plant sensing and communication into perspective and to have a project that would engage me once my kids left the house. I thank my chair, Mike Parrella, for giving me the time to work on this, and my wife, Mikaela Huntzinger, for encouraging me and for many discussions about plant behaviors and a million other things. Other people also influenced my views on this subject, most notably my colleagues in animal behavior at UC Davis, Gail Patricelli, Andy Sih, Judy Stamps, and Louie Yang. This manuscript was improved by Anurag Agrawal, Mikaela Huntzinger, Graeme Ruxton, and John Thompson. Christie Henry made the publication process seem effortless.

1 Plant Behavior and Communication

1.1 Plants and animals are different but also similar

We all learn intuitively that plants are not like us in many ways. Ask a child about the differences between plants and animals and the answer won't be about photosynthesis. They'll say, "Plants can't move" or "Plants don't do anything." It is true that plants don't appear to do the things that many of us find most interesting about humans and other animals—moving, communicating with one another, and displaying a great diversity of sophisticated behaviors that depend upon the particular situations in which they find themselves. But this intuition about plants is incorrect; plants sense many aspects of their abiotic and biotic environments and respond with a variety of plastic morphologies and behaviors that are often adaptive. In addition, plants communicate, signaling to remote organs within an individual, eavesdropping on neighboring individuals, and exchanging information with other organisms ranging from other plants to microbes to animals.

Plants lack central nervous systems; the mechanisms coordinating plant sensing, behavior, and communication are quite different from the systems that accomplish similar tasks in animals. The challenges that face plants are similar to those facing animals—finding resources, avoiding predators, pathogens, and abiotic stresses, acquiring mates, placing offspring in situations where they are likely to be successful. The modes of selection are also basically similar for plants and animals. As a result, natural selection has led to the evolution of solutions to these challenges that are often analogous. Nonetheless, there are many important differences between plants and animals that have led to very different adaptations, including the behaviors that will be considered in this book.

Although the issues that plants and animals face have similarities, their habits, abilities, and circumstances tend to be different in many ways. Most plants are capable of producing their own food, allowing them to spend much of their lives as factories converting resources (light, water, CO_2) into organic tissues. Since these resources are rapidly renewable, vegetative organs of many plants move relatively short distances and remain rooted in the soil. Higher plants are constructed of repeated modular units (leaves, branches, roots) that are far less specialized than the organs of higher animals (White 1984) . Many important processes are carried out by plant organs that are less centralized than their counterparts in higher animals. For example, plants lack a central nervous system, and consequently, phenotypic expression is determined locally in many cases. Plants acquire resources from many different organs, above ground and below, thus avoiding restrictions that would be imposed by one or a small number of mouths. Diverse plant structures arise from undifferentiated meristems that have the potential to produce any cell type, including germ cells and somatic cells. In addition, these diverse plant organs can be produced repeatedly during the lifespan of an individual. Important plant organs are generally found in multiple, redundant copies, making any single organ more expendable than similar organs in most animals. This open-ended growth form allows plants enormous developmental flexibility, an important attribute that was recognized by the Greek botanist Theophrastus approximately 300 years BC (White 1984, Herrera 2009). Developmental flexibility allows plants to respond to environmental cues and change morphology, adding or shedding organs in response to current or anticipated conditions and allowing plants to "forage" for light, water, and soil nutrients and to allocate resources to reproduction, growth, or storage.

The philosopher Michael Marder (2012, 2013) has recently introduced the idea that plants sense their environments by focusing attention towards some cues more than others. The attention is dynamic, allowing plants to selectively respond to shifting, current stimuli. Many studies of plant responses to resource heterogeneity attest to the ability of plants to sense many stimuli (chapter 2) and selectively respond (e.g., chapters 5–8). In addition, different cues vie for the attention of sense organs or receptors that are distributed throughout the plant's tissues. Plant sensing is not contemplative, but active, and translates into various behaviors (see section 1.2.1). Unlike memory, which is also displayed by plants but is biased towards past events, plant attention as described by Marder is focused on current stimuli.

As humans, we have a tendency to compare plants to humans and other higher animals. Such comparisons can be useful at times, since our awareness and understanding of human behavior and communication is so much better developed. However, such comparisons can range from counterproductive to absurd, if done uncritically. For example, several authors assert that plants can appreciate and benefit from hearing particular kinds of music, most famously in the popular book, *The Secret Lives of Plants* (Tompkins and Bird 1973). The hypothesis that plants may respond to music is not itself absurd, since plants can sense and respond to electromagnetic radiation and acoustic energy (Telewski 2006, Gagliano et al. 2012b). However, asserting that plants benefit from music without carefully controlled experiments is not science and has hurt progress in this field and acceptance of these ideas. In summary, while plants don't appear at first glance to behave, in some cases they have evolved functions that are analogous to those in animals, but with different mechanisms and capabilities.

1.2   Working definitions

1.2.1   Plant behavior

Before attempting to determine whether plants exhibit behaviors including communication, it seems reasonable to agree upon a set of criteria that define these phenomena. This is more difficult than it sounds. Although animal behavior is a relatively mature field, practitioners of that field cannot reach a consensus about what constitutes behavior (Levitis et al. 2009). Early behaviorists defined the term quite restrictively; Tinbergen (1955), for example, wrote that behavior included "the total movements made by an intact animal." This definition excludes plants (and other taxa) and it also fails to include inactivity or decisions to not reproduce as behavior, as well as changes in traits not involving physical movement. Some behaviorists wish to differentiate between intentional, purposeful behaviors from actions that result as unintended consequences of other processes. This distinction has proven to be problematic since it is virtually impossible to determine an animal's intent. A recent survey of behavioral biologists found little concordance about what phenomena could be considered behaviors although approximately half of the respondents identified plant responses to light as a behavior (Levitis et al. 2009).

The possibility that plants exhibit behavior is not a new suggestion. Charles Darwin's grandfather, Erasmus Darwin (1794:107), speculated that "vegetable life seems to possess an organ of sense to distinguish varying degrees of moisture, another of light, another of touch, and finally another analogous to our sense of smell." As is often the case in biology, Charles Darwin, who seems to have foreshadowed much of modern biology, provided detailed descriptions of many plant species that moved in response to light, gravity, and contact (Darwin 1880). In a more recent attempt to explicitly define behavior to include plants, Jonathan Silvertown and Deborah Gordon (1989) described behavior as a response to an event or environmental change during the course of the lifetime of an individual. Responses ultimately are the result of physiological changes that have a biochemical basis. Behavior differs from other physiological and biochemical reactions by occurring rapidly relative to the lifespan of the individual and requires a response to a stimulus. Furthermore, behavioral responses need not be permanent, and can be reversed if the stimulus changes. For example, the decision to expand a shoot into a sunny patch is reversible in the sense that it can be stopped and additional resources allocated to other tissues should that shoot become shaded. However, the resources that have been allocated to that shoot cannot be fully recovered. This definition of behavior does not include changes that are the result of ontogeny (Silvertown and Gordon 1989, Silvertown 1998). For example, the changes that occur as a seed germinates and expands its cotyledons and then its true leaves are not considered behavioral responses since they are part of a developmental program that is not plastic, once initiated. This definition of behavior is similar to one used by plant biologists to describe phenotypic plasticity (Bradshaw 1965), and behavior may be considered a form of plasticity that occurs rapidly and reversibly in response to a stimulus.

1.2.2 Plant sensing, eavesdropping, communication, cues and signals

Communication can be considered a behavior that provides information from a sender to a receiver. Communication also provides information that can cause the receiver of that information to respond (behave). As was the case for behavior, there is no agreed-upon definition of what constitutes communication either for animals or for plants (Scott-Phillips 2008, Schenk and Seabloom 2010). Most definitions require that receivers respond to cues or stimuli (Karban 2008). This requirement is considered necessary but not sufficient by most workers who study animal behavior since it includes situations in which receivers respond to cues from their abiotic environment. In keeping with accepted definitions, I will regard responses to stimuli as examples of plants sensing cues but not communicating. How plants sense their environments is fascinating in its own right and will be discussed at length later in this book. I will restrict my use of the term "communication" to situations in which emission or display of a cue is plastic and the response of the receiver is conditional on receiving the cue. For example, a plant that always attains a short compact growth form because of its genes is not responding to cues in a proximate, short-term sense. A plant that adjusts its morphology depending upon the cues that it receives from its neighbors may or may not be considered to be communicating.

Definitions of communication tend to emphasize either the exchange of information from a sender to a receiver (Smith 1977, Hauser 1996) or the requirement that the transfer of information be favored by natural selection (Maynard Smith and Harper 1995, Scott-Phillips 2008). Cues provide the receiver with accurate estimates of the relative probabilities of alternative conditions. Both the amount of information and its value to the receiver can be quantified, although doing so in a meaningful way can be challenging (Wilson 1975, Bradbury and Vehrencamp 1998).

Some authors consider communication to have occurred if information has been transferred that elicits a response in the receiver without regard to benefits, while others require that the sender, the receiver, or both benefit from the exchange (Wilson 1975, Bradbury and Vehrencamp 1998, Maynard Smith and Harper 2003). They distinguish between cues that do not necessarily benefit the sender and signals that increase the sender's fitness. The term "signal" is reserved for those situations in which exchange of information is beneficial for both the sender and the receiver. Receivers that respond to cues (as opposed to signals) may eavesdrop on the sender or may be engaging in communication with the sender. For example, an herbivore that locates its host plant by the volatile cues that the plant emits is eavesdropping on cues that the plant emitted for some purpose other than attracting herbivores. According to the authors cited above, "true communication" occurs when providing information in the form of a signal is not accidental but benefits the sender. True communication can occur when signals are transferred between cells, organs within an individual, or different individuals. Some authors require that the signal must have evolved because of the effects that it causes and therefore that both the sender and the receiver must experience a benefit by communicating (Maynard Smith and Harper 2003, Scott-Phillips 2008). This definition has many advantages and can explain the evolution of refinements to effective signaling.

The various definitions of communication can make reading this literature confusing. I have summarized the requirements associated with various terms in table 1.1. There is a consensus that communication occurs only when the signal is sent, is received, and causes a response (fig. 1.1). Communication requires that all three steps be present, although this book will consider each of the steps independently since plants sense and respond to environmental cues even when the cue was not intentionally sent by a living organism. One problem with the definition is that a signal that is missed by a receiver may be identical to one that causes a response. Is it a signal in one case but not in the other? This problem can be fairly easily resolved by stipulating that a signal will cause a response on average (Scott-Phillips 2008). A more serious problem with an adaptationist definition that explicitly relies on establishing that signaling arose because it benefited the sender and the receiver is that determining why a trait evolved is extremely difficult (Gould and Lewontin 1979, Endler 1986). For example, pigments make flower petals visible to insect pollinators, suggesting that they may be considered signals (Fineblum and Rausher 1997, Gronquist et al. 2001). However, these same pigments also deter floral herbivores, suggesting that they may have evolved for this purpose and that they should be considered cues, not signals, in communication with pollinators. This makes identifying true communication with any certainty almost impossible in many instances.

TABLE 1.1 Characteristics that define the phenomena considered in this book.

Phenomenon	Receiver responds to informative cue?	Cue is plastic & response is conditional?	Signal benefits sender?	Signal benefits sender & receiver?
Sensing	yes	?	?	?
Eavesdropping	yes	yes	no	no
Communication	yes	yes	yes	?
True communication	yes	yes	yes	yes

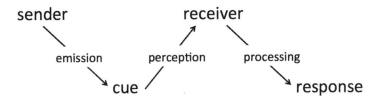

FIGURE 1.1 Communication occurs when a sender emits a cue (or signal) that is perceived (sensed) by a receiver. The receiver processes the information in the cue and responds.

1.3 Plant sensing and communication—organization of this book

In this book we will consider all three of the steps shown in fig. 1.1: what are the cues that plants emit, what are the cues that plants respond to, and how do plants or the organisms with which they are communicating change as a result? Although any single step is not sufficient for communication, each step can be fascinating regardless of what we call it.

Fig. 1.1 is a graphical representation of the scope of this book, and it will be used repeatedly to show the relationships between the various sections. Plants have considerable sensory capabilities, as receivers of various animate and inanimate cues (chapter 2). Sensitivity to cues can be influenced by past experiences, and plant responses are considered as learning in chapter 3. The properties of different cues that are involved with communication are examined in chapter 4, as are the mechanisms by which plants receive cues and emit them. Plants sense and respond to heterogeneity in resources (chapter 5) and herbivores (chapter 6). Communication between plants and animal visitors affect plant reproduction—particularly pollination and seed dispersal (chapter 7). Diverse interactions between plants and microbes are discussed in chapter 8. Sensing and communication affect plant fitness and can drive macroevolutionary patterns (chapter 9). An understanding of plant sensing and communication can lead to many useful application, considered in chapter 10.

2 Plant Sensory Capabilities

2.1 Plants sense their environments

Plants can sense many qualities about their environments. In this chapter, we will consider some of the environmental qualities that plants sense: light, chemicals, touch and gravity, temperature, sound, and electromagnetic forces. They do this using a variety of receptors and feedback mechanisms, including phytochrome receptors to detect light, mechanical sensing to detect gravity, and chemical feedbacks to detect CO_2. The stimuli that plants sense include both abiotic factors and those caused by other plants, microbes, and animals. Because we are more familiar with the ability of animals to detect their environments, those abilities will be compared to the sensory capabilities of plants.

Plants are also affected by their previous experiences; plant learning and memory will be discussed in chapter 3. Chapter 4 will explore what we know about the cues and signals used by plants to acquire sensory information and the cues that they produce that other organisms sense and respond to.

Plants live in a diversity of habitats: from deserts to rainforests, rooted in soil and free floating in water, under full sun or in full shade. These conditions often change over short spatial scales such that the seeds from a single mother may germinate in very different situations. Similarly, conditions may change rapidly so that an individual experiences a great range within its lifetime. As a result of this uncertainty and variability, plants have evolved the ability to sense their environments and to respond to their current and expected conditions.

2.2 Plants sense light

Plants require light to carry out photosynthesis, and light is often a limiting resource for them. Biologists have appreciated for centuries that plants respond to light gradients. We will first consider important properties of light that make it valuable as a resource for plants. Next we will consider light as a signal that indicates the presence of potentially competing neighbors. Finally, we will consider light as a source of visual information for animals.

2.2.1 Properties of light

Light is radiant electromagnetic energy; most biologically active light originates from nuclear fusion of hydrogen nuclei into helium on the sun. Light travels as a wave that can be described by its frequency, which is inversely proportional to its wavelength. Light from the sun includes a range of wavelengths, from cosmic rays (wavelengths less than 0.001 nm) to radio and slow electromagnetic waves (wavelengths of 1m–1000 km).

Visible light is in the middle of this spectrum, ranging from 390–700 nm. The atmosphere filters out most of the energy outside of the visible range by reflecting or absorbing it before it reaches the earth's surface.

Light behaves like a stream of packets of energy that are called quanta or photons in the visible range. Properties of light depend upon its mix of frequencies, which determine its spectral qualities (color). Another important property of light is its intensity, measured as the number of photons. Visible light shows some characteristics of shorter wavelengths (it can pass through objects to some extent) and some characteristics of longer wavelengths (it can bend around objects to some extent). Objects that are exposed to light energy from the sun undergo configurational changes because their outer electrons shift up to a higher energy level. Light in the visible range is sufficiently powerful to shift electrons and cause changes in the configurations of molecules (unlike weaker infrared radiation), but is not so powerful as to completely break chemical bonds (unlike stronger ultraviolet radiation). Objects are visible to animals when they reflect some of the light that reaches them. Light in the visible range is readily reflected by organic objects, unlike wavelengths that are larger or smaller.

Living organisms detect light when receptor molecules absorb electromagnetic radiation. Receptors absorb the radiation only when the frequency of the incoming energy precisely matches that needed to push the receptor molecule into its higher oscillatory state. As a result, a molecule can absorb only a limited subset of frequencies. Once energy is absorbed or trapped, it causes changes in the receptor that trigger a cascade of other changes in photosynthetically active plant tissues or in the nerve cells of animals. Energy that is absorbed by an organic molecule is subsequently lost through molecular collisions as heat.

When light reaches a surface, some is absorbed and some is reflected or scattered. A surface that reflects all wavelengths equally appears white while one that absorbs all wavelengths appears black. Surfaces that appear to have color are selectively absorbing some wavelengths and reflecting others. For example, leaves selectively absorb red wavelengths and reflect green; red wavelengths are absorbed by chlorophyll as it produces carbohydrates from CO_2 and water.

2.2.2   Light receptors of plants

Plants rely on three types of receptors that perceive light—phytochromes, cryptochromes, and phototropins (Smith 2000). The molecular structures and modes of action are known for all three although the phytochromes are the best understood. *Arabidopsis* has five different phytochrome genes of which phytochrome B is the most important (Sharrock and Clark 2002). Phytochrome genes have been found in all green plants including algae as well as certain bacteria. Phytochromes switch between an active and an inactive form when they are stimulated by light of particular wavelengths. One form (P_r) of phytochrome B has maximum absorption of red light and another form (P_{fr}) has maximum absorption of far-red light. Incoming sunlight has roughly equal proportions of red and far-red frequencies. However, green photosynthetically active pigments selectively absorb red light. For instance, chlorophyll a, found in higher plants, has a sharp absorption peak at approximately 665 nm, removing light in the red frequencies (Wolken 1995). The light that is reflected or scattered off of green foliage has a much greater proportion of far-red light than of red light; the ratio of red to far-red light under a dense leaf canopy is approximately 0.1 (Ballare 1999). In fact, the ratio of red:far-red is a very reliable predictor of shading, particularly shading by other foliage (fig. 2.1).

The P_r form of phytochrome B has an absorbance peak at 665 nm, which

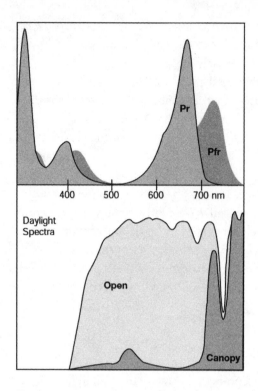

FIGURE 2.1 The top panel shows the wavelengths of light absorbed by phytochrome photoreceptors. The form P_r absorbs red wavelengths (650–670 nm) and the form P_{fr} absorbs far-red wavelengths (705–740 nm) most strongly. The bottom panel shows wavelengths of light in unshaded open situations and light from beneath a shaded plant canopy. From Smith 2000.

corresponds to the frequency of red light (fig. 2.1). The P_r form is biologically inactive, but when it absorbs red photons it is converted to the biologically active P_{fr} form (fig. 2.2). The active P_{fr} form has an absorbance peak at 730 nm, corresponding to the frequency of far-red light (fig 2.1). When it absorbs far-red light it converts to the inactive P_r form (fig. 2.2). Sunlight has a higher proportion of far-red light at dusk and plants adjust their sensitivity to the ratio of red:far-red light based on the time of day by means of a circadian clock (Vandenbussche et al. 2005).

After it converts to the active P_{fr} form, some of the cytoplasmic pool of phytochrome B travels into the nucleus, where it regulates gene expression, causing a cascade of chemical and physiological changes (Vandenbussche et al. 2005). The N-terminal half of the phytochrome molecule senses light and the C-terminal half regulates downstream transduction pathways (Smith 2000).

Phytochrome-mediated responses to light quality (red:far-red) can be extremely rapid because they involve the destruction of negative regulators (Huq 2006). In other words, the response system is always "ready to go" but may be temporally inactivated by light signals. Phytochromes D and E act

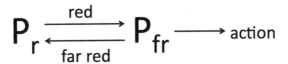

FIGURE 2.2 The P_r form is biologically inactive. When it absorbs red photons, it is converted to the active far form P_{fr}. The active form P_{fr} is converted back to the inactive P_r when it absorbs far red photons.

in a manner that is redundant with phytochrome B, suppressing the shade avoidance responses under conditions of full sun (high red:far-red) (Ballare 2009). Phytochrome A appears to act as an antagonist of phytochrome B to provide redundant control of the plant's response to shading.

The photosynthetic pigments in foliage also absorb blue light, although it is absorbed less selectively than red light. Phytochromes are involved in perception of blue light along with two different types of photoreceptors that are less well understood: cryptochromes and phototropins. Cryptochromes are flavoproteins found in both plants and animals; the light-sensitive chromophore of these enzymes is reduced by light, after which they move to the cell nucleus, where they cause stem elongation (Cashmore et al. 1999). Phototropins are flavoproteins that respond to blue and UV-A light (Briggs and Christie 2002). The N-terminal half acts as a light receptor and the C-terminal half contains a kinase that regulates other reactions. Phototropins in the cell membrane are stimulated by light to unfold and undergo phosphorylation that ultimately controls many responses including stem bending towards low-intensity light and away from high-intensity light, stomatal opening, migration of chloroplasts within cells, and solar tracking of leaves (Briggs and Christie 2002). The blue light receptors are thought to enable plants to assess the quantity of light (photon flux) in addition to its quality (spectral balance). In total, 14 light receptors have now been identified, some of which allow shoots to grow towards sources of light, others of which allow leaves and roots to avoid exposure to damaging UV (Rizzini et al. 2011, Briggs and Lin 2012, Yokawa et al. 2013).

2.2.3 Light receptors of animals

Although animals sense light for many reasons, here we consider light receptors in animals because animals respond to visual cues produced by plants. The diversity of receptors influences how animals perceive plants. Like plants, animals have cryptochrome receptors that are light sensitive and maintain circadian rhythms. Animals lack the phytochromes of plants, and cryptochrome is not the most important of their light receptors. Instead, they rely on rhodopsin, a highly conserved G protein–coupled receptor found throughout the animal kingdom (Wolken 1995). Rhodopsin is made up of a small molecule called retinal produced from vitamin A, a dietary beta-carotene. Retinal is a chromophore, which means that it traps incoming photons by elevating electrons to a higher energy state, similar to phytochromes of plants. Retinal is attached to a relatively large protein, called opsin, which makes up the bulk of rhodopsin. When retinal is activated by light, it changes from being a bent *cis*-isomer to a straight *trans*-isomer. This causes a cascade of conformational changes in the attached opsin. Rhodopsin is located in light-sensitive receptor cells of eyes that have very different shapes in different animals. Photoreceptor cells tend to be densely packed in a layer called the retina that forms a synapse with one or more nerve cells. When light activates and straightens the configuration of rhodopsin molecules, this opens ion channels in the receptor cell's plasma membrane that generate a nerve response. The receptors then slip back to their inactive form, releasing the energy as heat.

Rhodopsin absorbs green light most strongly. Slight changes in the chemical composition of the opsin portion of the molecule change the light frequencies that it responds to. Some animals also have droplets of colored oil in their photoreceptor cells that act to filter and change the absorption peaks. Animals that have receptors with sensitivities for at least two different light frequencies have the ability to perceive color. Humans have receptors with peak sensitivities for three frequencies: 450 nm corresponding to blue light, 530 nm corresponding to green light, and 560 nm corresponding to yellow light. When photons are absorbed by the receptors, they respond with a voltage change that it proportional to the amount of light they receive. These graded voltage changes, corresponding to light containing mixes of different frequencies, are passed on to the brain, which interprets them as color.

It is important to recognize that different animal species perceive visual inputs very differently. This diversity comes about at numerous stages in the vision process. First, different animals have receptors that absorb light at varying frequencies. Mammals that live in relatively low-light environments tend to have photoreceptors that concentrate on perceiving spatial patterns and differences in brightness. Their receptors have a single absorption peak, and they see the world as various shades of gray (Bradbury and Vehrencamp 1998). Some animals have more than three light receptors, including birds, reptiles, fish, and insects and other invertebrates that may

FIGURE 2.3 Many insects have photoreceptors that are sensitive to ultraviolet wavelengths. The black-and-white image on the top shows yellow monochromatic flowers of *Potentilla anserina* as they appear to us. The bottom image shows the same image as it would appear to an insect with photoreceptors that perceive ultraviolet light. Photograph courtesy of Bjorn Rorslett, naturfotograf.com.

have as many as twelve photoreceptor types that allow them to perceive light in the ultraviolet or infrared ranges (Cronin and Marshall 1989, Jacobs et al. 1991, Arikawa 2003). These capabilities make possible opportunities for sensing and communication that exceed our abilities. For example, insect visitors to many flowers perceive elaborate visual signals in the ultraviolet range called nectar guides that are invisible to humans (fig. 2.3).

Animals vary greatly in the structure of their eyes (Wolken 1995, Bradbury and Vehrencamp 1998). The simplest eyes are photoreceptors encased

FIGURE 2.4 The compound eyes of bees and other insects view the world differently than the camera eyes of vertebrates. Multiple ommatidia produce images that have relatively poor resolution and distortion due to the curvature of the hemispherical surface of the eye. The images on the top show what we see and the images on the bottom are output from a model that simulates how a honey bee perceives those shapes. Photographs courtesy of Andy Giger, andygiger.com.

by membranes found on algae, bacteria, fungi, and protozoa. These were followed in the phylogeny of life by "cup eyes" of many invertebrates with receptors lining the sides of the cup; the "pinhole eyes" provide information about the direction and intensity of light and are found on mollusks. "Camera-type eyes" found in hunting cephalopods and some vertebrates have a lens that can focus light to receptors and can produce a spatial map. "Compound eyes" found in insects evolved from cup eyes and are extremely sensitive to movement at the expense of fine-scale resolution. Eyes of higher vertebrates have two types of receptors: highly sensitive rods for low-light situations and cones with receptors for multiple frequencies that distinguish colors in bright light. As with different photoreceptors, the varied eye structures of different animals perceive the same objects very differently (fig. 2.4).

2.3 Chemical sensing

Many of the important interactions that plants engage in are carried out through chemical reactions. Plants must first sense and then acquire abiotic resources such as CO_2, water, and various solutes and minerals tied up in soil. Plants also sense and interact with other organisms, ranging from microbes to other plants to herbivores, pollinators, and seed dispersers, largely through chemical reactions. Recognition of these abiotic and biotic entities is often accomplished at a cellular level by plant receptors, proteins that span the plasma membrane. One end of the protein resides outside of the cell and can selectively bind to a chemical from the environment that provides the plant with information about the presence of an abiotic or biotic agent. The other end of the protein resides inside the cell and can trigger a cascade of reactions when binding with a chemical cue has occurred. Specific receptors of this sort allow plants to detect pathogens and abiotic stresses, among other things. In other cases, plants sense and respond to chemical changes in their environments without actual receptors but rather as the result of feedback caused by those chemical and physical conditions. Plant perception of CO_2 and water are examples that involve cellular feedback mechanisms rather than dedicated receptors.

2.3.1 Sensing resources

Photosynthesis requires both CO_2 and water, and either or both of these resources are commonly limiting for plants. Plants acquire CO_2 through pores in the leaf surfaces called stomata. Unfortunately, opening stomata to acquire CO_2 also allows water to evaporate. The stomata are surrounded by pairs of guard cells that regulate whether they are open, permitting CO_2 to enter and water to leave, or closed, reducing gas exchange. Blue light photoreceptors (phototropins) respond to light intensity and cause the guard cells to pump H^+ ions across the plasma membrane and out of the cell (Kinoshita et al. 2001, Scott 2008). This creates a negative electrical charge inside the cell and leads to an inflow of positively charged K^+ ions. To balance the osmotic potential caused by the inflow of K^+ ions, water diffuses into the cell, swelling the guard cells and forcing them to open.

Guard cells respond to many other signals in addition to blue light, integrating the needs of the plant for increasing CO_2 to enable photosynthesis while diminishing water stress (Assmann 1993). Elevated intercellular CO_2 concentrations cause stomata to close while lower concentrations stimulate the opposite. This feedback allows stomata to respond to the needs of the leaf, since photosynthesis diminishes the CO_2 concentration within the leaf and respiration in the dark leads to increased concentrations (Nobel 2009). Stomata also respond to plant water status, closing when the relative humidity of the ambient air falls or when leaf water potential decreases. Guard cells sense the relative humidity of the surrounding air and the water status of other cells including those in the root tip, responses that are mediated by several plant hormones (Roelfsema and Hedrich 2005). Abscisic acid (ABA) is produced in response to drought stress and causes stomatal closure (Assmann 1993). Cytosolic receptors on the surface of guard cells perceive ABA signals, which depolarize the plasma membrane and trigger the release of K^+ and other anions (Geiger et al. 2010). These drag water out of the guard cells, reducing turgor pressure and causing stomatal pores to close. Stomatal opening is not as well understood as closing. Light receptors in the plasma membrane trigger large negative membrane potentials, uptake of K^+, causing the guard cells to swell and the stomatal pores to open (Roelfsema and Hedrich 2005). In summary, plants appear to regulate their CO_2 and water budgets based largely on diffuse feedback and signaling systems rather than specific dedicated receptors that sense their internal or external conditions.

There is abundant and compelling evidence that plants sense and respond to many other chemicals in their environments. Mineral nutrients required by plants are distributed patchily throughout the soil. For at least a century, plant biologists have recognized that roots proliferate in those patches that contain higher concentrations of nutrients (Weaver 1926). Experiments confirmed the hypothesis that roots grew selectively in those richer patches (Drew et al. 1973, Drew 1975). Roots are able to sense and respond to the environmental conditions that they experience (see section 5.1.2) (Metlen et al. 2009, Cahill and McNickle 2011). However, the physiological details of how they detect those chemicals are still lacking, although incomplete pieces of the overall process are beginning to emerge. For example, *Arabidopsis* grows lateral roots when it encounters a patch of soil rich in nitrate. A nitrate transporter protein, NRT1.1, located in the plasma membrane of root cells represses lateral root production at low nitrate concentrations by promoting export of auxin (a growth stimulant) from these roots (Krouk et al. 2010). At high nitrate concentrations, export of auxin is inhibited and lateral roots proliferate. Exactly how the NRT1.1 protein recognizes and binds nitrate is still not known.

Arabidopsis plants grown under P-deficient conditions rapidly activated several genes in the roots that were associated with alleviating P starvation (Muller et al. 2004). Some of the genes involved in sensing other environmental chemicals and their locations in plant cells and tissues have also now been identified. A gene required for sensing moisture gradients has been identified in the root cap of *Arabidopsis* (Kobayashi et al. 2007).

2.3.2 Sensing hormones

Plants sense and respond to a small number of hormones that act as internal and exogenous signals to control many diverse processes. Hormones are active at extremely low concentrations. Unlike animals, plants do not possess specific glands that produce hormones; instead, plant hormones can be produced by multiple cells. Phytohormones can effect multiple responses depending upon their concentrations, the plant tissues with which they interact, and the environment. However, for plant hormones to be effective, they must trigger specific reactions in specific cells. There is also recent evidence that plants may detect and respond to hormones used by their herbivores (Helms et al. 2013).

2.3.3 Sensing pathogens and herbivores

The ability to distinguish self from nonself is highly conserved and allows even primitive organisms to defend themselves against invaders. Plants recognize the chemical profiles of pathogens and herbivores as nonself and respond to those signals. We know more about how plants detect pathogenic microorganisms, although there appear to be similarities with recognizing herbivores. Two different mechanisms have been described for sensing microbes based on chemical recognition—these differ in their specificity and when they are thought to occur in the evolutionary ploy versus counterploy of plant-microbe interactions (fig. 2.5).

Plants are able to recognize the basic patterns of entire groups of microbes; these patterns are referred to as MAMPs (microbe-associated molecular patterns) (Bent and Mackey 2007). These highly conserved portions of the microbe's genome are indispensable, in that they cannot be easily sacrificed or changed by selection without seriously impairing the microbe's abilities to colonize, survive, or reproduce. Plant genes responsible for detecting MAMPs have also been found to be heritable and evolutionarily conserved (Nurnberger et al. 2004). Plants can recognize many kinds of MAMPs including oligosaccharides, peptides, and enzymes from the pathogens, as

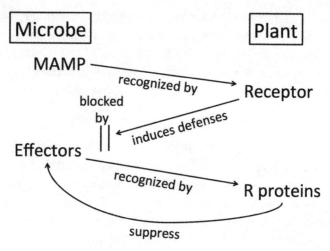

FIGURE 2.5 A model of communication between microbes and plants. Microbes possess MAMPs (microbe-associated molecular patterns) that are highly conserved and recognized by plant receptors. When plants recognize microbes, they induce defenses against the microbes that prevent infection. Microbes may contain effectors that block these induced defenses and allow the infection process to continue. Specific effectors may be recognized by R proteins of the plants, which may successfully suppress the microbial effectors, allowing plant-induced defenses to operate. This model is based on Jones and Dangl 2006 and Bent and Mackey 2007.

well as plant-derived products that indicate that they have been attacked. One well-studied MAMP is the highly conserved flagellin polypeptide, a constituent of many bacteria (Zipfel and Felix 2005).

All known MAMP receptors span the plasma membrane. We don't know exactly what these receptors look like although they appear to have an extracellular leucine-rich repeat structure and an intracellular protein kinase (Gomez-Gomez and Boller 2000). Following binding with the MAMP, the kinase portion of the receptor initiates reactions that provide defense in most cases. Pathogens that are unable to block perception of MAMPs fail to colonize (Bent and Mackey 2007). This has placed strong selection on the microbes to hide from plant surveillance systems by evolving "effectors" that allow them to avoid plant detection or circumvent defenses. For example, some bacteria have evolved flagellin proteins that are just different enough in their amino acid sequences to allow them to avoid detection even though their flagella are still functional. Others are thought to shed their flagella immediately after entering the host so that they become more difficult to detect.

Plants possess a second perception system that is far more specific and that recognizes, and binds to, particular microbe "effectors" (Bent and

Mackey 2007). These plant receptors are called R proteins (R for resistance), and they may act directly by binding with intracellular effectors or indirectly by binding with other plant proteins that have been perturbed and which provide a reliable cue of microbial action (fig. 2.5). In both cases, binding triggers protein activity and often leads to resistance. R proteins are structurally diverse but most have a leucine-rich repeat domain that determines pathogen specificity. In some cases, the N-terminal domain may mediate the initial binding of the R protein to the pathogen elicitor. Recognition and responses to pathogens are considered again in section 8.2.

Plants are thought to recognize herbivores in a manner similar to perception of pathogens, and several HAMPs (herbivore-associated molecular patterns) that plants recognize have been identified (Felton and Tumlinson 2008, Mithofer and Boland 2008). Plants appear capable of detecting chemicals released by insects walking over leaf surfaces, and secretions associated with oviposition and insect feeding. A receptor has been identified for one of these insect-associated patterns from caterpillar regurgitant (Truitt et al. 2004), although receptors are not presently known for other HAMPs.

Plant perception of HAMPs and other reliable cues of insects are not well understood. When some insect herbivores walk across a leaf surface, tobacco and soybean plants respond with a rapid, localized release of reactive oxygen species (Bown et al. 2002). This was followed by synthesis and accumulation of GABA (4-amino butyric acid), a nonprotein amino acid that may disrupt insect physiological functioning. Early instar tobacco budworm caterpillars failed to trigger the responses although heavier late instar caterpillars that walked for longer periods of time elicited GABA responses. These results are remarkable in their own right because they demonstrate exquisite sensitivity. Subsequent studies suggested that the plants were responding to slight disruptions in the plant epidermis caused by tarsal claws or proleg crochets of the caterpillars rather than to chemical cues associated with the herbivores (Hall et al. 2004).

Plants respond to mechanical wounding associated with chewing herbivores although responses to artificial damage in some plants have been found to differ from, and in others to be similar to, from responses to real herbivory. For example, many phenolic compounds accumulate in response to herbivory, mechanical damage, pathogen infection, and other stresses (Coleman and Jones 1991). However, on birch leaves, different amounts and types of phenolics accumulated depending on the type of damage that was inflicted (Hartley and Lawton 1991). Chewing by caterpillars induced greater levels of phenolics than leaf mining or mechanical cutting, and caterpillar oral secretions intensified the effects of mechanical cutting. By using a "mechanical robotic caterpillar" both the duration of feeding and the area of the leaf that was damaged were found to affect the response of lima bean plants (Mithofer et al. 2005). When the robotic caterpillar fed in a manner that matched a real herbivore, the plant produced an emission of volatile compounds that was qualitatively similar to that produced in response to real herbivory.

We have known for some time that plants respond to insect eggs by forming a localized necrotic zone around the eggs (Shapiro and DeVay 1987). More recently, *Arabidopsis* has been found to respond to oviposition by upregulating many genes associated with the production of the necrotic zone (Little et al. 2007). Plants are able to detect several chemicals associated specifically with insect oviposition (Hilker and Meiners 2010). Bruchins are esters that are produced in accessory gland secretions of weevils when they glue eggs on to legume pods (Doss et al. 2000). Plants recognize bruchins and develop callous tissue beneath the eggs that prevents weevil larvae from getting inside the pod. When female cabbage white butterflies place eggs on *Brassica oleracea*, they inadvertently allow benzyl cyanide to contact the leaf surface at the same time. Males place the benzyl cyanide during mating to reduce the likelihood that the female will remate. It offers the female no known benefit. When the benzyl cyanide contacts the leaf surface of this host plant, it causes changes in leaf chemistry that provide indirect plant defense (Fatouros et al. 2008). A growing number of plant species are being found that recognize chemicals associated with insect oviposition that cause these plants to change, although the potential consequences of these changes are not well understood. At present, the receptors have not been identified for these oviposition-associated elicitors.

Chemicals found in insect oral secretions or regurgitant are recognized by plants and are powerful elicitors of induced plant defenses (Felton and Tumlinson 2008, Mithofer and Boland 2008). Oral secretions from insects contain compounds that are specific to those herbivores; several proteins from caterpillar oral secretions and one peptide elicitor have been described that cause defensive reactions from plants. Low–molecular weight fatty acid conjugates are found in the regurgitant of several caterpillars, crickets, and fruit flies that are strong elicitors. Oral secretions from grasshoppers contain fatty acids that are also recognized by plants. Caterpillars synthesize HAMPs from precursors obtained from the host plants on which they are feeding. In maize, a protein located on the plasma membrane binds to the elicitor from caterpillar oral secretions and functions as a specific receptor for caterpillar regurgitant (Truitt et al. 2004). Plant hormones produced by other signals of herbivory upregulate this protein receptor, which may enhance plant perception of caterpillars.

Other early events in the perception of caterpillar oral secretions involve depolarization of the plasma membrane and formation of ion channels across membranes, although the cause-and-effect relationships involving these changes are not fully understood (Maffei et al. 2004, Maischak et al. 2007). Plant responses to herbivory are considered in more detail in chapter 6.

2.4   Mechanical sensing—touch

Plants respond to touch in a variety of circumstances (Braam 2005). Several of these reactions are widespread but fairly inconspicuous. For example, growing roots of many plants avoid obstacles (Darwin 1880). Stems that are exposed to repeated experimental touch or wind develop a shorter, sturdier growth form (fig. 2.6) (Biddington 1986). The stigmas and stamens of many species are sensitive to touch and move to increase the likelihood of successful outcrossing (Simons 1992). Some authors have argued that plant behaviors tend to be much slower than movements exhibited by animals (Harper 1985, Silvertown and Gordon 1989). This is true in many cases; however, the fastest motion yet observed in biology is the release of pollen by white mulberry flowers at velocities exceeding half the speed of sound (Taylor et al. 2006).

Other touch-sensitive responses are spectacular although less widespread. Carnivorous plants snap closed modified leaves to entrap insects that have touched several trigger hairs (Darwin 1893, Forterre et al. 2005). Species of "sensitive" plants, such as *Mimosa pudica*, fold up their leaflets and leaves rapidly (within milliseconds to seconds) in response to touch (Braam 2005). Further stimulation causes the response to spread to other leaves on the plant although cessation of stimulation causes the leaves to reopen gradually (seconds to minutes). Vines use a variety of organs that respond to touching a supportive substrate. These can be tendrils, including specialized shoots (e.g., *Vitis*), leaves (e.g., *Campis*), or inflorescences (e.g., *Passiflora*), as well as clinging roots (e.g., *Hedera*) and twining stems (e.g., *Ipomoea*). Some tendrils have been found to have greater levels of sensitivity than do humans (Braam 2005).

We do not have a complete understanding of how plants sense touch although several responses at both a subcellular and a whole-plant level have

FIGURE 2.6 The effects of repeated touch on the growth of *Arabidopsis*. The plants on the right were touched two times each day while those on the left were untouched controls. Touch caused plants to delay flowering and the produce shorter inflorescences. From Braam 2005.

been described (Chehab et al. 2009). These diverse phenomena may all share a conserved mechanism of response to mechanical signals at the interface where the plant cell wall connects to the plasma membrane (Telewski 2006). Within seconds after mechanical stimulation or stretching of the plasma membrane, there is a change in action potential and electrical resistance. This activates ion channels in the membrane that cause an increase in Ca^{2+} concentrations within the cell. These first steps are also observed when plants respond to mechanical strain, caused either by their own growth or by external forces (Braam 2005). The membranes themselves may respond to strain by stretch-activated changes in ion fluxes; alternatively, proteins that link the membranes to the cell wall may sense the strain and initiate ion fluctuations, as has been proposed for animal cells. In either case, Ca^{2+} fluxes cause upregulation of various genes over the 10 to 30 minutes following the initial stimulation. Genetic results indicate that at least 2.5% of the entire genome of *Arabidopsis* is upregulated in these responses (Lee et al. 2005b).

Despite these advances, there are no known mechanoresponsive receptors for plants (Monshausen and Gilroy 2009). Recent evidence implicates the plant hormone jasmonic acid (JA) in responses to touch (Chehab et al. 2012). *Arabidopsis* plants that were experimentally touched accumulated JA, while several mutants that were incapable of JA signaling failed to exhibit responses to mechanostimulation. Other signaling mechanisms are also involved. Mechanical stimulation in *Mimosa pudica* (sensitive plant) and Venus flytraps, is converted into an electrical signal (action potential) (Volkov et al. 2010). Voltage-gated ion channels then cause leaf movements; the entire process can be stimulated experimentally by applying low voltages.

Several diverse phenomena that appear to be responses to stimuli other than touch may be related to the touch responses and are sensed by plant membranes in a manner similar to that just described. At least some of the shade avoidance responses that some plants exhibit appear to involve responses to leaf touching that precede those stimulated by light quality (red:far-red ratios) (de Wit et al. 2012). When leaf tips of *Arabidopsis* continuously touch leaves of neighboring plants, the same "touch" genes that respond when membranes are mechanically stimulated were upregulated (Braam 2005). These changes in gene expression occurred before phytochrome signaling and were observed in mutants that were unable to respond to light (de Wit et al. 2012). In plants like *Arabidopsis* that form low-growing mats of rosette leaves, this response caused those leaves responding to touch to become slightly elevated so that they escaped the shade cast by neighbors more rapidly. The generality of this response for other species or growth forms is not known.

Mechanoreceptors associated with cell membranes also allow plants to sense cellular water status (Telewski 2006). The plasma membrane controls cell turgor by regulating the flow of water and solutes in or out of the cell. Changes in cellular turgor pressure can affect the overall morphology and mechanical properties of plant tissues. Stretch-activated Ca^{2+} channels in the plasma membrane or at the interface between the membrane and cell wall sense and respond to hypotonic and hypertonic conditions, allowing cells to maintain a favorable water balance (Hayashi et al. 2006).

Plants may use receptors on the surface of foliage to detect the presence of herbivorous insects. As insects walk across leaf surfaces, they rupture glandular trichomes that release plant hormones, including jasmonic acid, that rapidly activate the expression of genes associated with plant resistance (Peiffer et al. 2009). The pressure of "insect footsteps" alone was sufficient to cause these responses and preceded mechanical damage to plant tissue or deposition of herbivore saliva or oral secretions.

A similar mechanism is probably involved when plants sense gravity. It has long been observed that roots grow down and shoots grow up, regardless of the orientation of the germinating seed or developing plant (Knight 1806). Most models to explain how plants sense gravity rely on sensing a mechanical signal at the interface between the plasma membrane and the cell wall (Telewski 2006). Cells that sense gravity contain relatively dense starch granules, called statoliths (Morita 2010). Statoliths place more mechanical pressure on the plasma membrane on the bottom of the cell than on the top and this imbalance may result in strain and ion fluxes as described above. Statoliths are located immediately behind the root cap in roots and in the endodermal cell layer in shoots. Starch does not appear to be necessary for a statolith to sense gravity, although some dense particles that move in the direction of gravity do appear to be required. Once cells containing statoliths or an equivalent sensitive structure perceive a directional gravitational signal, that signal is converted into a biochemical cascade that ultimately results in asymmetrical auxin distribution and directed growth.

2.5 Plant sensing of temperature, electricity, and sound

It is not controversial at this point that plants can sense light, chemicals, and mechanical stimuli. It is possible, and perhaps even likely, that plants can respond to other stimuli that we are still unaware of. The popular literature is full of such reports and some replicated, controlled studies support these hypotheses, although they have yet to be fully examined and explained in peer-reviewed journals. More carefully conducted experiments also suggest that plants use senses other than those that are currently well understood. For example in one study, proximity to a neighboring plant influenced seed germination and growth of seedlings even when cues associated with light, chemicals, or touch were blocked (Gagliano et al. 2012a). The mechanism was not identified and opens the tantalizing possibility that other forms of sensing may occur in plants.

Exposure to cold stress changes many plant traits although it is controversial which of these changes represent plant responses rather than unavoidable damage (Scott 2008). Some well-documented responses to cold involve changes in gene expression and are clearly not simply the result of tissue damage (e.g., Artus et al. 1996). Plants damaged by chilling have compromised membranes; plants that are more sensitive to chilling damage have membranes with a lower ratio of unsaturated fatty acids to saturated fatty acids. Temperature alters the rigidity of cell membranes, and this dramatically affects the structure and functioning of those cells (Chinnusamy et al. 2007). The plasma membrane is a rigid gel at low temperatures, a liquid crystal at warmer temperatures, and a fluid liquid at high temperatures (Chinnusamy et al. 2010). Temperature also influences the conformational structure of membrane proteins, affecting Ca^{2+} fluxes and metabolic reactions. However, no specific cold receptor has yet been identified in plants.

Plants that are exposed gradually to cold temperatures can become acclimated by changing their cell membranes and reprogramming their transcriptome, which increases their tolerance to chilling (Thomashow 1999, Lee et al. 2005a). A nuclear factor, called *ICE1*, that suppresses expression of genes involved in cold and freezing tolerance when conditions are warmer, has been found to act as a master regulator in *Arabidopsis*. Other regulators are also involved in sensing cold although their modes of action are yet to be determined (Chinnusamy et al. 2010).

These responses to cold are quite general. In addition, a few plants have inflorescences that are able to thermoregulate, producing more heat as external temperatures fall (Nagy et al. 1972, Knutson 1974). Several maintain a relatively constant temperature that is independent of ambient air temperature (Seymour 2004). Production of heat in the inflorescences of thermoregulating species is controlled by changes in mitochondrial respiration; respiration rates as high as those of a hummingbird in flight have been recorded. These results suggest that plants are able to sense temperature and regulate respiratory rates accordingly. Although temperature receptors have not yet been found, a model has been proposed that involves control of temperature by adjusting concentrations of an important membrane protein (Wagner et al. 2008). Work in the future is likely to reveal much more about the mechanisms behind thermoregulation and temperature sensing more generally.

Both plant and animal cells use electrical signals to transmit information and to regulate many important physiological functions. Electrical signals generate action potentials, in which the electrical potential of a cell membrane rapidly rises and falls (Beilby 2007). Voltage-gated ion channels are closed when the membrane potential is similar to the resting potential inside the cell, but they open when the membrane potential rises, allowing rapid exchange of charged ions such as Cl^- and Ca^{2+}. These usher in a variety of other changes that have been described in many of the mechanisms used by plants to sense their environments described in this chapter.

Action potentials are clearly involved in many cellular processes in both plants and animals. In animals, action potentials stimulate nerves and muscles; some marine animals can generate large electrical charges that are used to capture prey, communicate, and monitor their environments (Bradbury and Vehrencamp 1998). A greater number and diversity of animals can detect electrical signals than can generate them, and many vertebrates have electric receptors that resemble pressure receptors and sound receptors.

Plants also produce and respond to electrical signals although plant biologists have been slow to investigate the functions of electrical signals and mechanisms responsible for generating and perceiving them (Davies 2004). Electrical signals move more rapidly (in the range of cm per second) over larger distances than other modes of information transfer (Fromm and Lautner 2007). When cell membranes are depolarized, action potentials are created that can propagate to other cells. Electrical signals in plants had been investigated in response to transient pulses caused by flames, heat, or herbivory, and recent results indicate that they are also involved in perception of light, touch, wounding, water status, and temperature changes, among other things. This area of research has until recently been viewed with skepticism by many plant biologists. It is likely to accelerate quickly in the near future now that most biologists recognize that electrical signaling is common and important for plants.

Surface potential changes that followed mechanical wounding in *Arabidopsis* correlated with accumulations of jasmonate and associated regulators (Mousavi et al. 2013). Blocking the electrical signal blocked the plant's ability to mount typical jasmonate-mediated responses. Injecting current into leaves induced jasmonate accumulations that followed the spatial and temporal pattern of actual herbivore damage. Examinations of plants that either overexpressed or showed reduced glutamate-like receptor activity suggested that plants may rely on these receptors to recognize pathogens and mechanical wounding (Kang et al. 2006, Bonaventure et al. 2007, Kwaaitaal et al. 2011, Mousavi et al. 2013). It is interesting that these receptors are functionally similar to the glutamate receptors that mediate neural communication, memory formation, and learning in vertebrate nervous systems.

Claims that plants hear and respond to music have been circulating for centuries (Tompkins and Bird 1973, Weinberger and Graefe 1973). Like light, sound propagates through a medium as a wave. Sound is produced when molecules of the medium are forcibly concentrated, causing them to collide with adjacent molecules so that the pressure differential is propagated away from the source. There is no doubt that plants can produce sound, as for example when tension in a plant's water transport system is abruptly released. However, there is no convincing evidence that this sound production is intentional or beneficial for plants, rather than a consequence of damaging water stress (Kikuta et al. 1997, Gagliano 2013).

Although plants produce sound, it remains unclear whether they can respond to sound. Plants (along with solid inanimate objects) are certainly capable of absorbing sound waves. Exposure of seeds and young plants to ultrasound (pressure waves with a frequency outside the range of human hearing) causes changes in plant chemistry, germination rates, and root metabolism and growth (Telewski 2006, Rokhina et al. 2009, Gagliano et al. 2012b). In a recent study, *Arabidopsis* plants that were played recordings of vibrations caused by feeding caterpillars primed their defenses against actual attack (Appel and Cocroft 2014). These responses were specific; plants exposed to wind or insect singing did not exhibit these defensive responses. What is still not known is whether responses to sound are important in nature and how plants might perceive sound waves.

In general, we have a very incomplete view of the sensory capabilities of plants. These capabilities constrain the extent to which plants can perceive their environments and communicate with other organisms. For this reason, I will describe what we know although I wish to emphasize our lack of understanding about sensory receptors throughout this book. As our knowledge of these receptors increases in the years to come, we will gain a fuller understanding of the types of cues that plants are capable of responding to. At present we know that plants can perceive light, chemicals, touch, temperature, electrical impulses and sound. They use a mix of dedicated receptors and diffuse feedback systems to experience their external and internal environments.

3   Plant Learning and Memory

3.1   Do plants learn?

Plants are capable of storing information and learning from their past experiences. In this chapter we will look at evidence that previous experiences with light, chemicals, resources, pathogens, herbivores, touch, and cold affect the behaviors that plants exhibit. Plants retain a "memory" of some of these experiences the effects of which are manifested in subsequent generations.

One of the hallmarks of animal behavior is the ability of animals to learn. Behaviorists have difficulty agreeing on a single definition of "learning," although most of the various definitions converge on a change in behavior that is the result of prior experience (Grier and Burk 1992, Breed and Moore 2011, Schacter et al. 2011). For learning to occur, the animals must "remember" past experiences so that memories of these experiences inform future behaviors. Learning and memory start with sensory inputs, and animals exhibit different kinds of learning depending on the duration of time over which memories are stored.

"Sensory" or "electrical" memory involves sensory inputs that last on the order of milliseconds to seconds. Short-term memory lasts for seconds or minutes, and most animals (humans included) can remember only 6 to 7 chunks or items in short-term memory. Short-term memory involves protein kinases that synthesize proteins and cyclic AMP, which changes the kinases. Short-term memories can be shifted to longer-term memories of unlimited duration, and these also involve protein kinases, as well as involving transcription and the growth of synaptic connections among neurons.

In chapter 2, I described the abilities of plants to sense and respond to their environments. These abilities often increase during repeated stimulation events. What criteria can we use to evaluate whether these changes in sensing and responding represent learning on the part of plants? Fundamentally, my working definition of learning requires that past events cause chemical changes that influence the sensitivity, speed, or effectiveness of subsequent plant sensing and associated responses. This is functionally similar to the storage and retrieval of information that occurs in animal brains, although the plant organs and mechanisms involved are quite different. However, there are limits to the transferability of concepts regarding animal learning to plants. Animal behaviorists sometimes stipulate that learning occurs only when the change in behavior is caused by the animal's nervous system, a requirement that necessarily precludes learning by any organism without a nervous system. In addition, an animal that has once broken its femur may be more likely to break the same bone again. Since this doesn't involve the nervous system, it would not be considered learning according to the definition used by most behaviorists, but it does represent learning according to the working definition of learning that I am using. I recognize that arguing that a broken femur has "learned" misses important connotations of the term. It is a matter of controversy whether it is useful or counterproductive to call experiences that change plant behavior "learning" and "memory." However, their functional similarities to animal processes are remarkable and are the focus of this chapter.

The properties of sensing and responding that are changed by past experiences have different names when they are caused by different stimuli; they are called "acclimation" or "hardening" for sensing and responding to abiotic stresses, and "priming" for sensing and responding to pathogens and herbivores. "Conditioning" is applied to all categories of plant learning and "stress imprint" is synonymous with memory. It is unfortunate that the term "conditioning" has slightly different connotations for plant biologists than for animal behaviorists, who reserve this term for situations in which there is a learned association between a stimulus and an outcome. The term "conditioning" is used without this connotation by plant biologists (Karban 2008). This chapter considers the evidence that plants learn and respond differently, depending upon their experiences, for each of the senses that were considered in chapter 2.

3.2 Learning, memory, and light

Charles Darwin was interested in the light-sensitive responses of plants; he observed that previous light exposure conditioned the responses of cotyledons (Darwin 1880:459–461). In his early experiments with *Phalaris canariensis*, he found that the direction and amount that cotyledons bent towards a source of light depended on the amount of light that they had previously received. Plants that had previously been exposed to light were less responsive than those that had been kept in the dark. Plants with relatively little previous light exposure not only moved towards the light source more strongly but also remained in that position for a longer time.

Since Darwin's day, experiments of this sort have been repeated and modified numerous times. For example, maize seedlings that had been grown in the dark oriented themselves in the direction of blue light, a proxy for bright light, when it was offered (Nick and Schafer 1988). They wanted to determine if plants formed memories of strong light sources. When secondary sources of the blue light were later introduced, the plants maintained a "memory" of the direction from which the light was initially coming. As long as the time interval between the pulses exceeded 90 minutes, the seedlings oriented in the direction of the initial pulse (Nick and Schafer 1988). The authors interpreted this result as indicating the time required for the plant to form a stable spatial memory. Before 90 minutes, the plant was unable to access and respond to the previous information.

Plant growth processes are regulated, at least in part, by changes in light quality. For example, bud burst in birches (*Betula pendula*) in spring requires chilling and also a change in the light environment (Linkosalo and Lechowicz 2006). When conditions experienced by young trees were experimentally manipulated, plants responded to the ratio of light to dark and the ratio of red to far red at sunrise and sunset, and compared the current ratio to that "remembered" from previous days.

3.3 Learning, memory, and perception of chemicals, resources, pathogens, and herbivores

Plants growing in nature experience their neighbors as complex combinations of inputs that involve patches of light, nutritional resources, allelochemicals, and associated changes in the microbial environment. Plants integrate information about their current conditions with stored "memories" of conditions that they have experienced over their lifetimes. Past "memories" in this context are accumulated experiences that have altered the morphology and physiology of the plant. For example, the growth architecture of a clover branch was influenced by its current neighbors and also by the neighbors that it had interacted with over the past year (Turkington et al. 1991). Pea plants allocated more biomass to roots in environments that were increasing in levels of resources in preference to those that were decreasing or constant, even if the absolute levels were lower (Shemesh et al. 2010). This suggests an ability to differentiate between current resource levels and those experienced in the past.

Drought stress is one type of environmental condition that at least some plants "remember." Plants respond to drought stress by elevating cellular concentrations of Ca^{2+}, which are involved in the signaling that causes expression of stress-responsive genes and ultimately greater tolerance to the stressful conditions (Knight et al. 1997, Knight et al. 1998). *Arabidopsis* plants that had previous exposure to drought stress responded more strongly and effectively to subsequent drought stress than plants without drought exposure. Responses to one stress may limit responses to others; plants with previous experience with oxidative stress responded less effectively to drought stress than plants without (Knight et al. 1998). *Arabidopsis* plants that were previously exposed to high concentrations of abscisic acid (associated with drought stress) became less responsive to light signals (Goh et al. 2003).

Agriculturalists take advantage of plant memory to improve the physiological state of some crop plants. A common practice used to improve crop establishment in saline soils is to prime the seeds by soaking them in salt solutions (Bradford 1986). Using this technique, seeds are allowed to imbibe water for a short period so that metabolic activity increases, but then the imbibing is stopped. Exposure to salt solutions allows the seeds to begin imbibing but prevents germination. Such seeds show dramatically increased germination and seedling development when they subsequently imbibe water under stressful saline conditions. For example, wheat seeds that had been soaked in various salt solutions experienced greater establishment and yield and reduced shoot concentrations of Na^+ than seeds that had not been primed when both were grown under stressful saline conditions (Iqbal and Ashraf 2007). Conditioning of seeds with salt solutions produced various physiological changes in tomato plants grown under salt stress, and these effects became more marked as the plants matured (Cayuela et al. 1996).

Plants often recognize pathogens and herbivores by their chemical signatures (see section 2.3.3), and there is considerable evidence that plant responses to these attackers can be primed by previous experience. Early experiments established that plants with previous exposure to pathogens were able to form hypersensitive necrotic zones that contained attackers more quickly and more successfully than naïve plants (reviewed by Chester 1933, Kuc 1995, Durrant and Dong 2004). Since those pioneering experiments, it has become clear that plants respond generally to necrotizing pathogens by inducing systemic acquired resistance (SAR), which protects them against the microbes that initially attacked and also against many other pathogens (Ryals et al. 1996, Sticher et al. 1997). Accumulation of the hormone salicylic acid (SA) is required for effective SAR (Malamy et al. 1990, Raskin 1992). Leaves that were exposed to pathogens contained increased pools of SA or its precursors (Chong et al. 2001). Such pools speed the local accumulation of SA after subsequent attacks, enhancing the effectiveness of the plant response.

Although SAR is a very widespread and well-studied process, the role of priming was not immediately appreciated since defensive responses became apparent only following a second attack (Conrath et al. 2006). Recently, several modes of priming have been identified in plants that exhibit SAR. *Arabidopsis* plants that experienced pathogen attack had higher levels of mRNA and inactive proteins associated with mitogen-activated protein kinases (MAP kinases) (Beckers et al. 2009). MAP kinases are found in all eukaryotes and are involved generally in transducing external signals from sensors to intracellular responses (Ichimura et al. 2002). Following a second challenge attack by pathogens, these MAP kinases showed elevated activity in primed plants and were associated with more effective resistance (Beckers et al. 2009). Plants that have been treated with commercial elicitors of resistance to pathogens (see section 10.3) also showed increased accumulations of these MAP kinases, suggesting that priming is responsible for the effectiveness of these elicitors.

Although SA is required for plants to show SAR, levels of SA accumulation did not necessarily predict the effectiveness of disease resistance in some instances (Cameron et al. 1999). *Arabidopsis* plants that accumulated azelaic acid prior to attack were able to mount a faster and stronger defense response than those that did not (Jung et al. 2009). Azelaic acid does not directly inhibit pathogens but primes plants to rapidly accumulate SA and to express genes associated with SA following an attack.

Another form of priming that protects plants against pathogens does not require SA and is termed induced systemic resistance. It involves the rapid stimulation of defenses that are associated with other plant hormones, jasmonic acid (JA) and ethylene (Van der Ent et al. 2009b). Soilborne microbes such as rhizobacteria and mycorrhizal fungi prime plants to respond to pathogens (chapter 8), although the mechanisms that cause these JA-mediated defenses have not been elucidated yet. Plant recognition of microbe-associated molecular patterns (MAMPs) has been identified as an early step in this process (Van der Ent et al. 2009b). The microbes that prime plants protected by this form of resistance do not cause expression of target genes until the plants are actually infected by pathogens, although they are believed to cause slight increases in regulatory transcription factors that are sufficient to prime defensive genes (Van der Ent et al. 2009a). Defenses against herbivores often involve the JA and ethylene pathways and microbial priming associated with induced systemic resistance also protects plants against some insect herbivores (Zehnder et al. 1997, Zehnder et al. 2001, Van Oosten et al. 2008).

Although we have only been aware of priming in response to herbivores as a phenomenon for around a decade (Engleberth et al. 2004), examples have quickly accumulated that involve many different defenses. Frequently these studies have compared plants that were exposed to volatiles emitted by previously damaged plants and volatiles from undamaged control plants. Herbivore-induced volatiles cause plants to accumulate jasmonic acid (JA), a plant hormone that regulates defenses against many chewing herbivores (Engleberth et al. 2004, Frost et al. 2008). In many cases, JA increases only after subsequent attacks to the primed plant. For example, oral secretions of caterpillars induced a transient burst in JA that declined to baseline levels within approximately 2 hours (Stork et al. 2009). A second elicitation with caterpillar oral secretions suppressed the JA burst but subsequent elicitations caused more rapid JA accumulation. Furthermore, the baseline levels of JA to which the plant returned increased with each additional elicitation experience (Stork et al. 2009).

In response to a second attack, plants primed with herbivore-induced volatiles upregulated various inducible genes related to plant defense (Engleberth et al. 2004, Kessler et al. 2006, Engleberth et al. 2007, Ton et al. 2007, Frost et al. 2008). These changes allowed plants to respond more rapidly and more effectively to herbivore attack, reducing herbivore feeding and performance in many cases (Kessler et al. 2006, Ruuhola et al. 2007, Ton et al. 2007, Rodriguez-Saona et al. 2009). In several instances, the responses made plants more attractive or rewarding to predators and parasites of their herbivores, and plants that had been primed reacted more effectively than those that were naïve (Choh et al. 2004, Choh and Takabayashi 2006, Heil and Silva Bueno 2007, Ton et al. 2007, Peng et al. 2011).

Another event that can prime plants to heighten their reactivity is oviposition by herbivores. Tomato plants that were primed by oviposition by *Helicoverpa zea* induced more effective chemical defenses against the caterpillars once they began to feed (Kim et al. 2011). Sawfly larvae (*Diprion pini*) that fed on the twigs on which they had hatched from eggs grew less quickly, suffered greater mortality, and became less fecund adults than larvae that fed on twigs that had not received eggs (Beyaert et al. 2012). Insect eggs are an unusually reliable predictor of impending herbivore risk although the specific cues that plants perceive are not known at this time.

Most of the examples of priming by herbivores that have been described are relatively rapid and short-lived, lasting hours to days. However, priming may be more durable, lasting for the life of the plant. Tomato seeds can be primed by treating them with elicitors of resistance to herbivores and pathogens. Plants grown from these seeds were more resistant to a variety of herbivores and pathogens for at least 8 weeks, taking them into maturity (Worrall et al. 2012). In some instances priming may persist over even longer times, lasting for at least 5 years (Ruuhola et al. 2007). Birch trees that had been attacked 5 years earlier responded more strongly to a new challenge by caterpillars than trees that had not previously hosted caterpillars. Caterpillars that fed on trees that had a history of attack had lower pupal weights and developmental rates than those that fed on naïve trees. For trees that had been exposed to herbivores 5 years previously, hydrolysable tannins were associated with reduced pupal weights, whereas naïve trees exhibited no such negative relationship.

This example may be indicative of a widespread trend. Many cases of induced resistance were not detected until the season following herbivore exposure (Karban and Baldwin 1997: table 4.1). It is possible that many of these examples of "delayed induced resistance" may actually be cases of priming. There are also numerous examples of situations in which plants that were exposed to multiple bouts of herbivory showed stronger induced responses than plants that were exposed only once (Karban and Baldwin 1997:25). In these cases there is no simple way to distinguish between the effects of plant memory and the possibility that more accumulated damage induced a stronger or more rapid effect.

These examples may also be indicative of an ontogenetic trend. In general, plants are most inducible against herbivores during their juvenile stages (Karban and Baldwin 1997, Barton and Koricheva 2010). This ontogenetic trend may be caused by greater rates of growth during early development or because more tissue is undifferentiated. It may also reflect the fact that seeds and young seedlings have a different relationship with their herbivores than do older and larger plants. Seeds and seedlings are likely to be killed by herbivores, which can be thought to act as predators of these young plants. Larger plants are more likely to survive attacks by herbivores, which can be thought to act as parasites. The fitness of young plants may be more dependent on sensing their current and future environments and priming responses than that of older plants.

In these examples, priming generally makes plants more responsive to current conditions, which presumably increases fitness. Of course, priming caused by past experiences may also constrain their current abilities to respond plastically to current conditions. For example, *Abutilon theophrasti* seedlings responded to cues of shading (low red:far-red light) by exhibiting elongated stems (Weinig and Delph 2001). Seedlings that were elongated in this manner became less responsive to later light cues than plants that had not previously been primed. Similarly, young plants of *Solanum dulcamara* responded specifically to the different species of herbivores that first attacked them (Viswanathan et al. 2007). Following this initial response, they became less responsive to subsequent attacks and less plastic in their defensive phenotypes than were naïve plants of a similar age. Priming can be costly because it reduces the plant's ability to respond to other threats.

Three classes of mechanisms have been proposed to explain conditioning or priming by chemicals (Conrath et al. 2006, Bruce et al. 2007). When plants are first exposed to the conditioning agents they could produce and accumulate proteins that remain inactive but ready to be "hyperactivated" by a second exposure (Conrath et al. 2006). Conditioned plants could also accumulate a crucial transcription factor that becomes operational only following the second exposure but allows conditioned plants to act more rapidly. For example, transcription factor genes that are induced by drought and cold stress have been identified in *Arabidopsis* (Bruce et al. 2007). Priming could also cause epigenetic changes in the DNA activity without changing the nucleotide sequences (Bruce et al. 2007). The initial conditioning event could remove repressors so that the genes responsible for a particular response are kept in a potentially active or "permissive" state. Future work will be required to evaluate these potential mechanisms.

3.4 Learning, memory, and touch

In addition to recognizing specific chemicals associated with pathogens and herbivores, plants sense potential attackers by responding to touch receptors. There is growing evidence that previous experience with attackers primes plants to respond more effectively to physical stimuli, although our current understanding of these phenomena is incomplete. Physical contact by caterpillars or moths ruptured glandular trichomes of tomato plants and was sufficient to induce defensive genes (Peiffer et al. 2009). Insect feeding or application of methyl jasmonate (a form of JA associated with feeding) induced plants to produce new leaves with higher densities of glandular trichomes. We have historically assumed that trichomes increase resistance to herbivory but they may also increase sensitivity to herbivore touch. Touch can have similar effects as chemical cues. Plant responses were alike after receiving either physical contact by caterpillars walking across the leaf surface or exposure to green leaf volatiles emitted by plants whose leaves were crushed by feeding. These two stimuli are both indicative of herbivore attack, and both induced increased synthesis of GABA (Hall et al. 2004, Mirabella et al. 2008). The similarity in the plant responses to these two stimuli suggests that herbivore contact may prime plants to respond more effectively to other damage signals (Hilker and Meiners 2010). It is worth noting that GABA is involved in plant responses to stress, since it is a well-known neurotransmitter that controls synaptic transfers in human brains and may play a similar role in plant signaling in response to diverse stresses (Bouche and Fromm 2004). In addition, methyl jasmonate is involved in plant responses to touch (e.g., tendril coiling [Falkenstein et al. 1991]) as well as plant responses to herbivory.

Some plants respond to touch by folding their leaves. Leaves that have been closed in response to touch eventually reopen, although the timing of reopening varies and appears to exhibit some memory of past experiences. Experimental leaf damage increased the time to reopen for damaged *Mimosa* leaves but not for other adjacent leaves on the same plant (Cahill et al. 2013).

Other responses to touch, not associated with damage by herbivores, also exhibit short-term memories. For example, Venus flytraps require stimulation by two different hairs, before snapping closed and these hairs must be stimulated within 40 seconds of one another (Trewavas 2009). Individuals can be thought of as remembering the first stimulation for 40 seconds before discarding that information. Venus flytraps close in response to electrical stimulation, the purported signal, without mechanical stimulation of the trigger hairs (Volkov et al. 2008). Plants that are exposed to subthreshold charges can store these charges for up to 50 seconds and close once the sum of the small charges they have received exceeds the threshold value. Similarly, tendril coiling in pea plants requires two signals, blue light and mechanical stimulation. Either of these signals can be remembered for several hours (Jaffe and Shotwell 1980).

3.5 Learning, memory, and cold

Many plants can become acclimated to the cold, though the mechanisms of cold perception remain unknown (section 2.5 above). For example, during acclimation they acquire tolerance to freezing conditions after exposure to low but nonfreezing temperatures (Chinnusamy et al. 2007). Naïve rye was killed by temperatures of $-5°C$ but after experience with low nonfreezing temperatures, plants survived at temperatures as low as $-30°C$ (Thomashow 1999). During cold acclimation, plants profoundly change their cell and tissue structures and reprogram their metabolisms and gene expression patterns. The multiple responses to cold differed in different plant tissues, with reproductive tissues being the least cold tolerant (Chinnusamy et al. 2007). Tolerance to cold and freezing is associated with reduced growth and metabolism, and plants deacclimate after exposure to warmer temperatures (Guy 1990). Acclimation to cold is a much slower process, occurring over days, weeks, and months, than the rapid plant responses to heat shock and tolerance of hot temperatures. Deacclimation to cold tends to occur more rapidly than acclimation (Guy 1990).

Some plants also have a vernalization requirement—they will not produce sensitive reproductive tissues until they have experienced a species-specific number of hours of cold temperatures. Cold acclimation is most effective if plants can respond rapidly to the onset of cold, before their sensitive tissues become damaged. In contrast, vernalization is most effective if plants respond slowly, only after prolonged exposure to cold. Vernalization prevents plants from flowering or producing seeds in the autumn or winter, before the threat of damaging cold temperatures or freezing has passed. In some *Arabidopsis* genotypes flowering is normally prevented by a powerful repressor (Michaels and Amasino 2000). Prolonged cold temperatures inactivate this repressor, and it remains inactive throughout multiple mitotic divisions during spring growth. In many plants, the apical meristem is the sensitive organ, and vernalized meristems are capable of responding to other cues that trigger flowering. However, seeds of the next generation revert to the suppressed state and are incapable of flowering without cold. This requirement for vernalization prevents plants from producing reproductive tissues during times that will cause them to be ineffective or die.

Cold exposure alone does not cause flowering but does make the plant competent to flower if it receives other inductive cues (Amasino 2004). In a series of classic experiments, *Hyposcamus niger* required exposure to cold temperatures followed by photoperiodic cues (long days) that indicated the arrival of spring before flowering (summarized in Amasino 2004). Plants exposed to cold but grown under noninductive photoperiods failed to flower. Plants that were vernalized and then later exposed to inductive photoperiods flowered normally, suggesting memory. Vernalized meristems retain the ability to flower for up to at least 300 days in the absence of additional light cues (Taiz and Zeiger 2002).

The mechanisms that plants use to measure the duration of cold are not currently known. Prolonged cold uniquely induces genes in *Arabidopsis* that are required for the vernalization process (Amasino 2004). There does not appear to be a conserved vernalization pathway common to all plants; for example, cereals have a different repressor than *Arabidopsis* (Greenup et al. 2009).

3.5   Transgenerational memory

Vernalization of plants in one generation does not persist through the seeds into the next generation (Amasino 2004). The term "vernalization" was coined by the Russian geneticist Trofim Lysenko, who described the phenomenon in the first half of the 20th century. Lysenko fabricated results about vernalization and other Lamarkian-style inherited acquired characteristics. His goal was to demonstrate that favorable environmental conditions would permanently improve the genetic stock. Scientists who contradicted his views were coerced into silence or exiled to Siberian labor camps by the Soviets. Not surprisingly, Lysenko's fabricated data and censorship crippled the advancement of biology in the Soviet Union (Caspari and Marshak 1965). As a result, the knee-jerk reaction of Western biologists was to discount the entire hypothesis that events experienced by parents could influence the traits of their offspring. While there is no convincing evidence that such transgenerational influences can affect vernalization, recent work indicates that many traits can be inherited by epigenetic mechanisms (Jablonka and Raz 2009). The phenotype of the offspring is influenced both by the genes it inherits and by parental experiences without a genotypic change involving traditional inheritance of DNA sequence alleles. Epigenetic means of vernalization and other transgenerational changes can come about as the result of altered gene function caused by changing DNA methylation, histone modifications that affect the structure of the DNA strands, or related processes (Jablonka and Raz 2009).

Parental experiences can also influence offspring because those experiences determine the resources and other provisions that parents are able to offer (Rossiter 1996). For example, whether maternal plants were grown in sun or shade determined the life history strategies of their offspring (Galloway and Etterson 2007). Offspring that were manipulated to have the life history that was informed by the experiences of their mother were far more successful than offspring that were manipulated to have strategies that were uninformed. Epigenetic modifications can be caused by cues rather than by actual experiences, in what could be called transgenerational priming. For example, exposure to volatile cues associated with damage to neighbors caused demethylation of promoter regions associated with defensive genes in maize (Ali et al. 2013). This allowed those receiver plants to store "memories" of increased risk of herbivory, which primed their defenses for later attacks. Both epigenetic and maternal mechanisms can be favored by selection if cues that the parents perceive are predictive of the environments that their offspring are likely to experience (Bondurianksy and Day 2009, Holeski et al. 2012).

The first well-substantiated reports of transgenerational memory were published in the 1990s, and many similar papers have appeared more recently (table 3.1). As epigenetic mechanisms have become better accepted, more empiricists have found, or been willing or able to publish, results involving inherited environmental effects. The list of plant species for which transgenerational effects have been found is still relatively small, and, as is the case for many other mechanistic phenomena, *Arabidopsis* dominates these studies (table 3.1). However, parental experiences that have been found to cause transgenerational effects are diverse and include physical stresses (UV, salt, cold, and heat) and biotic stresses (competitors, pathogens, and herbivores). In most cases these past experiences prepare offspring to better handle similar challenges in the future (table 3.1).

The majority of studies to date have examined how well prepared offspring are in the generation that immediately follows the parents that experience the stress or attack. However, in a few cases, transgenerational memories have been observed persisting for at least 4 generations (Molinier

TABLE 3.1 Examples of transgenerational effects showing the species affected, the cue that elicited the effect, the response exhibited by individuals in subsequent generations, and the number of generations beyond the maternal generation for which a response was documented.

Author and year	Plant species	Cue	Response by offspring	gens beyond maternal
Roberts 1983	*Nicotiana tabacum*	TMV infection	+ virus resistance	2
Schmitt et al. 1992	*Plantago lanceolata*	light environment	seed weight and life history	1
Agrawal et al. 1999	*Raphanus raphanistrum*	herbivory	− caterpillar growth	1
Boyko et al. 2006	*Arabidopsis*	salt stress	+ tolerance to salt stress	1
Molinier et al. 2006	*Arabidopsis*	UV or flagellin	+ rate of recombination	4
Blodner et al. 2007	*Arabidopsis*	cold exposure	+ cold tolerance	1
Galloway & Etterson 2007	*Campanula americanum*	light environment	more appropriate life history	1
Holeski 2007	*Mimulus guttatus*	Damage	+ trichome density	1
Whittle et al. 2009	*Arabidopsis*	high temperature	+ heat tolerance	3
Kathiria et al. 2010	*Nicotiana tabacum*	TMV infection	+ resistance to infection	2
Luna et al 2012.	*Arabidopsis*	bacterial infection	+ resistance to infection	2
Rasmann et al. 2012	*Arabidopsis*	herbivory, JA	+ defense; − caterpillar growth	2
Rasmann et al. 2012	*Solanum lycopersicum*	herbivory, JA	− caterpillar growth	1
Slaughter et al. 2012	*Arabidopsis*	bacterial infections	+ resistance to infection	1
Verhoeven & van Gurp 2012	*Taraxicum officinale*	JA	− caterpillar feeding	1
terHorst & Lau 2012	*Lotus wrangelianus*	herbivory	− herbivore damage	1

et al. 2006). There is some suggestion that the extent of the parental experience may influence the longevity of the transgenerational response. For example, a single inoculation with an avirulent strain of *Pseudomonas syringae* conditioned *Arabidopsis* offspring in the following generation to increase resistance to infection but not in the subsequent generations (Slaughter et al. 2012). However, repeated inoculations with a virulent strain of the same pathogen produced transgenerational resistance that was maintained beyond a single generation (Luna et al. 2012).

In conclusion, there is abundant evidence that plants learn and have memory. These adaptations have the potential to contribute to plant fitness. These processes may provide missing links to explain unresolved patterns of inheritance of resistance to diseases and other traits (Holeski et al. 2012). Studying these processes has long been taboo for many scientists, either because of prejudice about the definitions of learning and memory or because of the sociology of Soviet science. However, we are in a renaissance period with regard to these topics, and these processes are likely to become more widely appreciated now that we are investing more energy in examining them.

4 Cues and Signals in Plant Communication

4.1 The nature of cues and signals

In this chapter we will consider the nature of cues and signals that plants use. Animal behaviorists differentiate between cues and signals (chapter 1). Cues are perceived by a receiver and cause a response. Cues are not necessarily intentionally emitted by the sender and may reduce the sender's fitness in some instances; cues may be generated by an abiotic source in other instances. In contrast, signals are emitted by organisms and have been shaped by natural selection such that they increase the sender's fitness on average (Otte 1974, Bradbury and Vehrencamp 1998, Maynard Smith and Harper 2003). Plants perceive environmental cues and respond to them. Plants also produce cues that are perceived by other organisms. Cues operate over different spatial scales, since communication occurs between individuals and also between tissues or cells of one individual.

One more necessary characteristic of both cues and signals is the ability of the receiver to control its physiological response to the stimulus (Schenk and Seabloom 2010). A plant cannot choose to ignore a source of intense heat or a toxin; while these both cause plant responses, they are not cues or signals. As with most criteria used in definitions, this requirement is less straightforward than it seems. For example, the roots of the invasive plant *Centaurea maculosa* exude catechin and probably other allelochemicals that inhibit the growth of neighboring plants (Bais et al. 2003, Blair et al. 2009), possibly by making soil nutrients less accessible (Thorpe et al. 2009). However, a few resistant species were found to excrete oxalate in response to exposure to catechin, effectively blocking its effects (Weir et al. 2006). Is it reasonable to consider catechin a cue in some of these situations, but not others?

In general, information is exchanged within and among cells and organs of a single individual by use of signals. These signals may move by diffusion or active processes and they may travel within the plant or move outside of the plant structure. In addition, plants acquire information-rich cues from other organisms and the abiotic environment that contain reliable information about current or future conditions. For both signals and cues, the plant must perceive a stimulus and respond physiologically.

The ability to sense the environment is useful to an organism only if it results in responses that are conditionally appropriate and increase the receiver's fitness (the right-hand side of fig 1.1). Since plants respond to a wide variety of visual, chemical, tactile, and electrical sensory stimuli (chapter 2), it follows that they may recognize a great diversity of cues and signals. In this chapter we will consider responses of plants to both environmental and internal cues. This chapter will also outline the kinds of cues that plants produce that cause other organisms to respond. The biochemical pathways that result in differential plant growth or in defense against pathogens and herbivores are remarkably complex and involve many signals at multiple control points. Rather than attempting to present a comprehensive list of the cues and signals that plants employ, I will describe several of the most important signals used to transfer information within cells, among cells, between plant individuals, and between plants and other organisms. Next, we will examine the cues and signals used by plants to perceive competitors, pathogens, and herbivores. Finally, this chapter will consider the cues and signals that animals use to perceive plants.

Signals and the pathways they trigger differ in their speed, which may be exceedingly rapid (within seconds) or considerably slower (years). Electrical impulses and changes in charged particles across plasma membranes cause signaling within seconds. Signals relying on existing proteins can occur within a few minutes. For example, plants use existing proteins to move chloroplasts out of harm's way when there is a dramatic change in exposure to high light levels. Blue light receptors in the plasma membrane stimulate kinase activity that causes a reorganization of the cytoskeleton, resulting in reorientation of the chloroplasts. When gene expression is involved, signaling tends to take minutes to hours. For example, it may take plant cells up to 30 minutes to transcribe a gene, process its mRNA, and export the mRNA to the cytoplasm. Then more time is required for the synthesis and intracellular movement of proteins. Vernalization and other complex responses may operate on a time scale of years.

Cues and signals that are involved in plant communication may be either fixed or plastic phenotypes. Cues that can change are interesting to us for a variety of reasons, although it is worth keeping in mind that many important cues are fixed (e.g., color or morphology). Plastic cues have the potential to provide more timely information because they change and hence can be more reliable. Plastic cues give plants the potential to respond conditionally to their changing environments. For example, the cues emitted in response to recent events (e.g., abiotic environment, attack by herbivores) allow plants to be more responsive and more interactive than fixed cues. Both the emission of plastic cues by plants and conditional plant responses to environmental cues are processes that fall under my definition of plant behavior.

All of the plant responses that have been examined in detail can be elicited by multiple cues that appear to have largely redundant effects. In addition, different cues and signals operate at different points in a plant's response. Since the cues and signals associated with the perception of any environmental situation do not operate independently but rather affect each other, I consider them as an interconnected and, in some cases, coordinated suite although the details of their interactions are often not well understood. Several properties of cues and signals seem relevant: where they occur, the events that stimulate them, their targets, their spatial range, the speed with which they propagate, and their specificity. The best-studied examples of suites of cues and signals are those that allow plants to avoid or moderate shade and engage in other forms of competition.

4.2   Plant competition—light and hormonal cues

Competition between plants is an important force that affects plant fitness and structures plant communities. Plants compete for many resources including light, water, and nutrients and respond to cues indicating the relative availability of these resources. These cues vary greatly in many regards including their location within the plant, the conditions that generate their emission, the tissues they target, the spatial range over which they are active, how quickly they move, and their specificity. Several important cues

TABLE 4.1. Some common cues used by plants to sense competitors

Cue	Location	Stimulus	Target	Range	Speed	Specificity
blue light flux	cell membranes	various	photoreceptors	meters	very fast	low
red & far red flux	stems	shading	phytochrome A?	meters	very fast	low
red : far red ratio	leaves	shading by leaves	phytochrome	meters	very fast	low
ethylene	ER of all cells	shading & stress	ethylene receptors	cms	fast	high
nitrate	root plasma membrane	low nitrate	nitrate receptors	cms	slow	low
root exudates	root hairs	competitors	varied	cms	slow	mostly high
auxin	all cells	shading, nutrients	varied	cms to ms	slow	high

that mediate competitive interactions between plants are summarized in table 4.1 and described in this section.

Plants respond to two kinds of light cues (Ballare 2009)—those indicating the quantity of light and those indicating its spectral quality (see section 2.2.1). Changes in the fluxes of blue, red, and far-red wavelengths, or of the total photosynthetically active radiation, can be detected. Plants also respond to the ratios of red to far-red light (Ballare 2009). Either the quantity or quality of light can act as a cue that signals to a plant tissue that it is being shaded by another photosynthetically active tissue. These cues can provide information about light availability at the scale of different organs within an individual and at the scale of individual plants. At a subcellular level, phytochrome B acts as a mobile signal, leaving the nucleus under conditions of low red:far-red that occur when a plant is being shaded. This subcellular migration allows expression of genes that ultimately results in stem elongation (Lorrain et al. 2008).

Ethylene, a gaseous hormone, provides a second signaling system indicating the presence of competitors (Pierik et al. 2006, Kegge and Pierik 2010). In dense stands of vegetation, ethylene concentrations build up in the canopy (up to 4-fold in greenhouse situations). Low red:far-red ratios incite ethylene synthesis (Finlayson et al. 1998, Pierik, Cuppens, et al. 2004). Under these conditions, ethylene can stimulate growth and trigger stem elongation. Ethylene controls growth in a complicated manner that is concentration-dependent, species-specific, and even tissue-specific. Ethylene also causes reallocation of plant resources, causing leaves that are not receiving light to senesce and redistributing those resources to leaves that are receiving more light (Pierik et al. 2006). Because ethylene is volatile, it can provide information (cues) about the competitive environment among neighboring plants (Kegge and Pierik 2010). However, ethylene provides little specific information about the identity of the neighbor or its strength as a competitor because it is produced by essentially all plants. Studies involving mutants that were insensitive to ethylene suggest that perception of this volatile may allow individuals to assess their own position relative to that of their plant neighbors. Insensitive tobacco plants displayed delayed detection of neighboring competitors and were outcompeted by wild-type plants that could respond to ethylene (Pierik et al. 2003). In addition, plants that were insensitive to ethylene lacked feedback on their own status and consequently emitted excessive quantities of ethylene (Bleecker et al. 1988, Knoester et al. 1998).

Several plant hormones act as additional signals that transfer information to other cells and tissues within a plant (Pierik et al. 2013). Auxin, gibberelic acid, and ethylene all accumulate in response to shading, promote the shade avoidance response, and have the ability to move readily within a plant (Taiz and Zeiger 2010). Auxin, in particular, is both a short-distance signal and a long-distance intraplant signal that operates between adjacent cells and between distant tissues and organs. Auxin promotes elongation in stems and inhibits elongation in roots. It generally moves from the shoot apex or root apex towards the base of the plant. This movement is independent of gravity and is mediated by carriers both into and out of cells (Li et al. 2005). In *Arabidopsis*, the flow of auxin is directed to meristematic tissues through the control of specific proteins (PIN3) located at the plasma membranes (Wisniewska et al. 2006, Keuskamp et al. 2010). Once in the meristematic tissues, auxin activates and derepresses a suite of responsive genes that ultimately control the extent and morphology of plant growth. In roots, auxin concentrations are controlled, at least in part, by transporter proteins, NRT1.1 in the plasma membranes (Krouk et al. 2010). These transporters inhibit root growth in unfavorable soil patches by moving auxin away from primordia that could give rise to lateral roots. In resource-rich patches the transporter proteins do not remove auxin, allowing it to accumulate and stimulate the growth of lateral roots (Krouk et al. 2010). Movement of auxin throughout the plant is regulated by a combination of many different factors; essentially all of the other plant hormones and signaling compounds affect rates of auxin transport and therefore growth (Taiz and Zeiger 2010).

4.3 Cues used in plant defense

Plants must recognize that they have been attacked by pathogens or herbivores quickly and accurately if they are to mount successful defenses. In some cases they recognize cues of impending damage such as insect footsteps and eggs (see section 2.3.3). Individual cells recognize the invader and then produce systemic signals alerting the entire plant and, in some instances, its neighbors. The cues and signals that plants use to defend against pathogens and herbivores and some of their properties are summarized in table 4.2.

4.3.1 Membrane potential (Vm)

Following attack, the earliest plant signal that is consistently detected is a change in the membrane potential of the plasma membrane. These propagate at a rate of approximately 1 cm/min (Maffei et al. 2007). Many stimuli can trigger depolarization of transmembrane potentials such as mechanical damage, H_2O_2 introduced by feeding herbivores or induced by the plant, oral secretions from herbivores, and cell wall fragments from invading pathogens, although the signature of the depolarization is specific to the stimulus (Zebelo and Maffei 2012).

Depolarization is followed by changes in the action potential of plasma membranes which can travel much faster and farther (Wildon et al. 1992, Maffei et al. 2007). The action potential is a transient signal transported through plasmodesmal networks and phloem tissue; it is poorly understood (Zebelo and Maffei 2012). Mechanical wounding of *Arabidopsis* stimulated plasma membrane depolarization that traveled along vascular connections at speeds as high as 13 cm/min (Mousavi et al. 2013). These action potentials could be stimulated experimentally by electrodes and correlated spatially and temporally with patterns of jasmonate activity throughout the plant. Current injection caused expression of many, though not all, of the genes that respond to damage, indicating that other signals are also involved.

TABLE 4.2 Some cues used in plant defense against pathogens and herbivores.

Cue	Location	Stimulus	Target	Range	Speed	Specificity
membrane potential (Vm)	plasma membrane	damage, oral secretions, pathogen cell walls, H_2O_2	plasma membrane	cms	slow	low
action potential	plasma membrane	damage, Vm, ion fluxes	plasma membrane	cms to ms	very fast	very low
Ca^{2+} flux	cytosol	herbivory, biotrophic pathogen	Ca^{2+} binding proteins	intracellular to cms	fast	high
reactive oxygen species (H_2O_2)	plasma membrane	pathogen, herbivore attack	varied	intracellular to cms	slow to fast	low
MAP kinases	nucleus	attacks, stresses	transcription factors	intracellular	fast	high
salicylic acid	cytoplasm	pathogens	NPR proteins signaling SAR	intracellular to interplant	azelaic acid, MeSA fast	high
jasmonic acid	chloroplast & peroxisome	herbivory, wounding	repressor proteins	intracellular to interplant	fast	high
ethylene	any tissue	herbivory & many other stresses, development	receptors on endoplasmic reticultum	intracellular to interplant	fast	high or low
green leaf volatiles	plastids and cytosol	tissue disruption	unknown	intracellular to interplant	fast	low

Action potentials have also been found to be very rapid signals involved in plant responses to heat stress and chilling.

4.3.2 Ca²⁺ flux

Electrical signals are followed closely by Ca^{2+} flow into the cytosol of cells that have been attacked (Maffei et al. 2007). Oral secretions from herbivores and biotrophic activity of pathogens were sufficient to cause Ca^{2+} signaling although mechanical wounding was not. Changes in Ca^{2+} concentrations cause additional depolarization of membranes and positive feedbacks with other signals that can amplify the message (Zebelo and Maffei 2012). Other ions such as Na^+, K^+, and Cl^- may accompany the changes in Ca^{2+} concentrations. Following attacks by pathogens or insects, Ca^{2+} is transported from the apoplast or organelles into the cytosol (Reddy et al. 2011). Different abiotic stresses and attacks by pathogens and herbivores cause specific Ca^{2+} signatures that vary in the number and timing of spikes; these signatures are thought to elicit specific and appropriate physiological responses to a given signal. Specificity is enhanced by a great diversity of Ca^{2+}-binding proteins that activate enzymes such as kinases and initiate cascades of defensive responses. Ca^{2+} fluxes occur over relatively small spatial scales and are most important for intracellular signaling.

4.3.3 Reactive oxygen species

Reactive oxygen species (ROS), such as hydrogen peroxide H_2O_2, superoxide O_2^-, and hydroxyl radicals OH, are produced rapidly after invasion by pathogens or herbivores and in response to many stresses (Lamb and Dixon 1997, Miller et al. 2009). Mechanical wounding may or may not elicit their production, depending upon the plant species. ROS can provide direct defense against the invaders when they accumulate at high concentrations and as signals at lower concentrations. They also trigger hypersensitive cell death, which kills the attacked cells but effectively traps the invading organisms or deprives them of nutrients. Ion fluxes associated with rapid changes in membrane potential regulate enzyme activity levels that control the production of defensive ROS (Wu and Baldwin 2009). However, Ca^{2+} fluxes appear to act both upstream and downstream of ROS in signaling pathways. ROS are produced relatively slowly following recognition of molecular patterns associated with pathogens or herbivores and move through the apoplast (Torres 2010). The ability of these small molecules to diffuse through membranes enhances their utility as signals and they may propagate systemically very quickly at rates up to 8.4 cm/min (Miller et al. 2009). They can oxidize regulatory proteins and alter transcription factors in the nucleus, and thereby affect gene expression. They have been found to have very diverse signaling roles in different contexts.

Plants possess another related signaling pathway that involves nitric oxide, NO, although its role is not as well understood (Klessig et al. 2000, Delledonne 2005). Activation of NO synthesis is triggered by increases in cytosolic calcium concentrations; NO also mobilizes intracellular Ca^{2+}. Both NO and ROS are required for some plants to mount a hypersensitive response to pathogen attacks. NO also induces the accumulation of later signaling hormones, specifically salicylic acid in response to pathogen attack and jasmonic acid in the case of herbivore attack.

4.3.4 MAP kinases

Mitogen-activated protein kinases are involved in the next step in transducing the signals from activated receptors to regulate plant responses to pathogens, herbivores and other stresses (Asai et al. 2002, Maffei et al. 2007, Rodriguez et al. 2010). These enzymes are widely conserved among all eukaryotes; they are inactive in their normal states but generally become activated by two phosphorylation events. Upon activation they mediate transcription of genes that provide defense. MAP kinases respond to oxidative bursts involving ROS and also regulate further ROS production (Rodriguez et al. 2010). Similarly, MAP kinases control the accumulation of, and otherwise interact with, the plant hormones involved in responses to stresses and attack such as ethylene, salicylic acid, jasmonic acid, and auxin.

4.3.5 Salicylic acid

Plants that are attacked by pathogens may successfully defend themselves by mounting a localized hypersensitive response that contains and kills the microbes at the site of infection. They also become more resistant to pathogens at distant sites as the result of systemic acquired resistance (SAR) (Ross 1961, Durrant and Dong 2004). Salicylic acid (SA) is the signaling molecule that triggers SAR (Gaffney et al. 1993). Concentrations of SA increase at the site of infection, systemically in tissues distant from the site of infection, and in the phloem sap (Malamy et al. 1990, Metraux et al. 1990). Crop plants can be inoculated with chemical analogues of SA, and these treatments are sufficient to induce resistance against a wide variety of pathogens (see section 10.3) (Gorlach et al. 1996, Lyon 2007). SA moves readily through the plant in the phloem sap, although removing damaged leaves before SA had time to build up in the petiole did not prevent the expression of SAR (Rasmussen et al. 1991). This result indicated that another, more rapid, mobile signal was also involved. Recently, azelaic acid was identified as this mobile signal in *Arabidopsis* (Jung et al. 2009). Plants treated with azelaic acid accumulated SA more rapidly following subsequent attack by pathogens. Ca^{2+} signaling, which binds a suppressor, also appears to be required to allow the synthesis and accumulation of SA (Du et al. 2009). SA activates a signaling protein (NPR), which migrates to the nucleus and enhances transcription of "pathogenesis-related" proteins. A great diversity of pathogenesis-related proteins are involved with providing plants with SAR (van Loon et al. 2006a). In addition to producing these proteins, plants with SAR are primed to respond more rapidly and more robustly to attacks, although the mechanisms responsible for this priming have not been resolved (Conrath 2009).

These signaling mechanisms allow plants to coordinate defenses against pathogens at the scale of single cells and systemically among distant cells within an individual. Plants also respond to volatile cues when distant tissues or even neighbors have been attacked by pathogens. Tobacco plants that were infested with tobacco mosaic virus emitted relatively large quantities of methyl salicylate (MeSA), the volatile methyl ester of SA (Shulaev et al. 1997). Plants that were exposed to methyl salicylate converted this to SA and produced pathogenesis-related proteins and lesions that prevented the spread of viruses. The generality and ecological relevance of communication between plants that leads to SAR is uncertain.

Recent work suggests that methyl salicylate (MeSA) may be a mobile signal that moves throughout tobacco plants to provide SAR (Park et al. 2007). Plants with diminished ability of attacked leaves to convert SA to MeSA at the site of infection or to convert MeSA back to SA at systemic tissues were not protected. The mechanisms involved in these events are not known, although it is likely that MeSA is perceived and bound by specific proteins that convert it to SA in receiver plants (Vlot et al. 2008). Work in this field on several different model plants suggests that SAR may also involve other mobile signals in addition to MeSA. MeSA has also been found to be repellent to some herbivores (Pettersson et al. 1994).

4.3.6   Jasmonic acid

Physical wounding or attack by herbivores stimulates the octadecanoid pathway, which results in an increase in jasmonic acid (JA), usually within 1–2 hours. Jasmonic acid has been found to regulate most of the diverse plant responses to herbivory as well as many other plant functions including defense against nonpathogenic microbes (Creelman and Mullet 1997). In addition to direct defenses, JA has also been found to mediate several "indirect defenses" against herbivores including emission of volatiles that attract the predators and parasites of herbivores and extrafloral nectar that nourishes those predators (see section 6.3) (Avdiushko et al. 1995, Dicke et al. 1999, Heil 2004). Mutants of *Arabidopsis* that were deficient in precursors of JA were susceptible to herbivores that wild-type plants were defended against (McConn et al. 1997). Similarly, wild tobacco plants with silenced JA pathways in the field were more susceptible to adapted herbivores and were also damaged by generalist herbivores that normally do not feed on this host (Kessler et al. 2004).

The precursors to JA are produced in the membranes of chloroplasts and peroxisomes; these precursors, along with MAP kinase activity, are believed to be the rate-limiting steps in the pathway (Wu and Baldwin 2009). Before JA can activate the expression of plant defenses, it must first be conjugated to isoleucine, an amino acid. This conjugate then binds to repressor proteins that inhibit transcription of defensive genes; binding to the conjugate ultimately leads to the degradation of the inhibitor (Chini et al. 2007, Thines et al. 2007). Released from these inhibitors, genes for defensive proteins and for healing plant wounds are transcribed (Johnson et al. 1989, Farmer and Ryan 1990, Creelman and Mullet 1997).

JA accumulations are greatest at the site of damage and spread 10–20 mm into undamaged tissue (Schulze et al. 2007). However, after herbivory by Colorado potato beetles or mechanical wounding, tomato and potato leaves systemically accumulated proteinase inhibitors that defended both the damaged leaves and undamaged leaves (Green and Ryan 1972). JA is transported through the phloem to target cells, where it activates defensive genes (Zhang and Baldwin 1997). Tomato mutants that were unable to either produce the jasmonate signal or were unable to recognize the jasmonate signal failed to activate those genes (Li et al. 2002). Thus, although the precise identity of the intercellular signal is still unknown, it is likely that either JA or a related compound from the octadecanoid pathway acts as an intercellular systemic signal.

Damaged leaves also emit volatile compounds that can trigger defenses in receiver tissues of the same and different individuals. Early lab experiments involving potted tomato plants incubated with clipped sagebrush branches indicated that the sagebrush emitted methyl jasmonate (MeJA), the volatile methyl ester of jasmonic acid (Farmer and Ryan 1990). MeJA induced synthesis of defensive compounds in the tomato plants, which contacted sagebrush only via shared air space. Native tobacco plants naturally occurring in the field near sagebrush became more resistant to their herbivores when the sagebrush was experimentally clipped (Karban et al. 2000), although there is some question whether MeJA concentrations were sufficient to serve as the volatile interplant cue (Preston et al. 2004). Airborne MeJA is taken up into leaves and rapidly converted to biologically active JA and other conjugates (Tamogami et al. 2008). Other related jasmonates, such as cis-jasmone, may also be involved as very rapid volatile cues or may be the actual signal, rather than MeJA (Bruce et al. 2008, Matthes et al. 2010).

4.3.7 Ethylene

Ethylene is another volatile hormone that often works in concert with jasmonates (O'Donnell et al. 1996). Ethylene synthesis is induced rapidly and transiently by wounding, herbivory, infection by pathogens, and other forms of stress (Williamson 1950, Abeles et al. 1992, van Loon et al. 2006b). Ethylene synthesis is also induced by experimental applications of jasmonates and conversely, ethylene can regulate endogenous levels of JA (O'Donnell et al. 1996, Arimura et al. 2002, von Dahl and Baldwin 2007). Ethylene synthesis can occur in virtually any cell in higher plants. Ethylene is a small volatile molecule that can diffuse rapidly through plant tissues and also move between the air and plant cells. Ethylene directs many diverse plant processes following perception by membrane-bound receptor proteins (see section 2.3.2). This leads to activation of histidine kinases and various transcription factors and ultimately to altered gene expression that mediates the different responses. JA and ethylene are required for many direct and indirect induced defenses against pathogens and herbivores, although the effects of ethylene in regulating defenses are context-dependent, and ethylene may fine tune responses to other hormones (van Loon et al. 2006b, von Dahl and Baldwin 2007). Ethylene affects many diverse processes and was discussed earlier in this chapter as a cue of plant competition.

Wounded plants release green leaf volatiles (GLVs), the C_6 compounds (aldehydes, alcohols, and esters) and terpenes (C_{10} monoterpenes, C_{15} sesquiterpenes, and C_{20} diterpenes) that we associate with the smell of newly cut grass (Matsui 2006). GLVs are emitted constitutively in low concentrations, but emissions increase greatly after disruption of tissue (Pare and Tumlinson 1999). GLVs are emitted very rapidly following damage, sooner than other compounds (Turlings et al. 1998). C_6 GLVs are synthesized via the octadecanoid pathway, similar to JA, primarily in plastids. Other terpenoids are produced via either the mevalonic pathway from acetyl-CoA in the cytosol or via the methylerythritol phosphate pathway from pyruvate in the plastids (Dudareva et al. 2004). These compounds are produced in epidermal cells, secretory structures, or glandular trichomes and emitted from these sites or from storage organs (Kant et al. 2009). Emission of GLVs occurred from undamaged leaves on wounded cotton plants, suggesting the involvement of another systemic signal (Rose et al. 1996). Emission of GLVs following damage also exhibited a diurnal pattern in several plant species, indicating that emissions were not simply caused by tissue rupture (Loughrin et al. 1994, Arimura et al. 2005). GLVs were adsorbed by damaged and nearby undamaged tissues, which may concentrate them under natural conditions, enhancing their effectiveness (Choh et al. 2004).

It is not known how plants perceive GLVs. Regardless, terpenoids that were induced by herbivory caused increased transcription of many genes involved in defense against herbivores and pathogens (Bate and Rothstein 1998, Arimura et al. 2000). Volatile terpenoids also caused the biosynthesis of ethylene and JA (Arimura et al. 2002). GLVs have the potential to act as volatile signals mediating communication between plants. Lima bean leaves exposed to GLVs from plants infested with spider mites accumulated gene products associated with defense (Arimura et al. 2000, 2001). In addition, GLVs primed plants to respond more rapidly in terms of producing JA when they later encountered herbivore regurgitant (Engleberth et al. 2004). GLVs also have direct bactericidal, fungicidal, and insecticidal properties and can repel ovipositing herbivores, independent of their role as signals that are recognized by plants (Gershenzon and Croteau 1991, De Moraes et al. 2001, Matsui 2006). Exposure to GLVs and terpenoids increased egg predation rates by generalist predators (Kessler and Baldwin 2001). Plants that were silenced in their production of these volatiles received more natural herbivory (Kessler et al. 2004). Naturally emitted GLVs were important for attracting parasitoids of herbivores (Shiojiri et al. 2006). Both herbivores and parasitoids have been found to have high sensitivities to detect C_6 volatiles and terpenoids (Kant et al. 2009).

4.3.9 Properties of cues and signals that make them useful for defense

The previous sections have considered the cues that plants perceive that indicate the presence of current or future attacks by pathogens and herbivores. They also included the signals that plants use to transmit this information within and among cells and individuals. Several properties will tend to make these cues and signals more useful.

To be effective, any cue must stimulate a receptor that the receiver possesses; this requirement for a receptor creates an important constraint on plant sensing. Evolution is expected to favor greater sensitivity on the part of receivers to reliable cues, but evolution is limited by the range of existing variation. In addition, useful cues and signals must travel from the stimulus or sender to the receiver, and this provides an additional set of constraints. For example, Ca^{2+} ions passively diffuse across plant membranes, but at relatively slow rates. They are still useful as cues and signals if there is some assisted process for traversing lipid membranes by regulating membrane potential, by active transport across the membrane, and by specific gated channels and pumps. Informative cues must be generated soon after attack and must travel quickly to receivers, be they adjacent cells, tissues, or plants. Lipophilic molecules are better able to pass through plant membranes, which are built of lipids. Most volatiles used as cues and signals are lipophilic; they are often made from hydrophilic precursors by removing the hydrophilic parts of the molecules and made more volatile by shortening their carbon skeletons (Dudareva et al. 2004). Small molecules pass through barriers like membranes more easily and are more volatile; thus, the compounds that are used as volatile cues and signals all tend to be small (Dudareva et al. 2004).

Despite these advantages, the two most important intraplant signals for defense, SA and JA, are not volatile and not particularly mobile. Both can move through the phloem, but this mode of transport imposes additional constraints. In some instances, the expression of induced systemic resistance has been limited to the tissues that share vascular connections, leaving other tissues vulnerable (Viswanathan and Thaler 2004, Orians 2005).

Both SA and JA can be converted to their methyl esters, which have much greater mobility both within and among plants.

4.4 Cues and signals emitted by plants that animals sense

Plants and animals interact in many diverse ways. Because animals eat plants, foliage and other valuable plant structures should be selected to have low apparency (detectability) by herbivores or low desirability once detected. Many plants are not capable of cross-fertilization without the aid of external forces to move pollen to receptive stigmas of different conspecific individuals. Plants often rely on animals for this service, and selection favors those reproductive structures that are attractive and rewarding to flower visitors that provide quality pollination. Similarly, animals are frequently effective vectors of seeds for plants, and plants attract and then reward animals that disperse seeds. Herbivory, pollination, and seed dispersal will be considered in chapters 6 and 7. This current section considers the plant cues and signals that animals perceive and respond to. These fall into two main categories: light cues used in visual communication and chemical cues used in olfactory and gustatory communication. These cues and some of their properties are summarized in table 4.3. Plants also engage in diverse interactions with microbes that attack plants, help plants acquire resources, interact with other organisms, etc. These will be considered in chapter 8; the cues and signals that mediate those plant-microbe interactions are poorly known and will be considered briefly in chapter 8 as well.

Different animals have different sensitivities to different cues and the match between cue and receiver will influence the effectiveness of communication. For example, many vertebrates including frugivorous birds can distinguish visual color patterns of flowers and fruits at distances of tens of meters (Schaefer et al. 2006). The same cues are visible to some insects, including foraging bees, at only a few centimeters (Spaethe et al. 2001). Most vertebrates and honeybees require high light intensities to discern colors (Somanathan et al. 2008). However, nocturnal bees and hawkmoths can distinguish color even by starlight on moonless nights. Anyone who has owned a dog is aware of our relatively inferior senses of smell and hearing. Most insects rely on chemical communication systems and have sensory abilities that are orders of magnitude more sensitive than those of mammals. It is difficult, if not impossible, to view cues and signals with anything other than a human perspective and unfortunately this perspective presents severe limitations.

TABLE 4.3 Plant cues and signals perceived by animals.

Cue	Location	Stimulus	Characteristic of stimulus	Target	Range	Specificity
chlorophyll	thylakoid membrane, most tissues	light	green	rhodopsin in eyes	cms to kms	low
carotenoids	thylakoid membrane	light	yellow, orange	rhodopsin in eyes	cms to 100m	low
anthocyanins	all tissues	light	red, purple, blue	rhodopsin in eyes	cms to ms	low to moderate
volatiles	epidermis & elsewhere	odors	greatly diverse	receptors	cms to kms	high
chemicals	all tissues	taste	sweet, umami, bitter, sour, salty	5 taste receptors	0 distance	high

4.4.1 Visual cues used by animals

Plant pigments provide visual cues that animals perceive. Pigments absorb certain wavelengths of light and transmit the remaining wavelengths. Structural colors formed by interference or scattering are commonly produced by animals but rarely by plants. Most plant colors are produced by chlorophylls, carotenoids, and anthocyanins (Taiz and Zeiger 2010, Schaefer and Ruxton 2011).

Chlorophylls are the photosynthetic plant pigments that capture light energy. They are bound to proteins on the thylakoid membrane of chloroplasts and are ubiquitous on aboveground plant tissues. Chlorophylls absorb mainly at the blue and red ends of the visible spectrum, reflecting green wavelengths. The green color of plants varies depending upon the density of chlorophyll pigments, the presence of other pigments, and other surface features such as waxes and trichomes. Reception of light by rhodopsin in animal eyes was described in section 2.2.3. Animals vary greatly in their abilities to perceive chlorophyll from distances ranging from centimeters to kilometers. Since chlorophyll is common to almost all plants, green color is not a specific cue, although herbivores have the ability to differentiate among subtle variations of green.

Carotenoids are terpenoids and, like chlorophyll, they are found in all photosynthetic organisms and absorb light energy. Carotenoids are also found on thylakoid membranes. When they are excited by light, they transfer that energy to chlorophyll for photosynthesis. They also protect the photosynthetic machinery from damage caused by excessive light by quenching the excited state of chlorophyll. Carotenoids absorb light most strongly in the violet, blue, and green wavelengths and reflect yellow and orange light. The yellow light reflected by carotenoids is often masked by green chlorophyll. When chlorophyll is less abundant, for instance when it is broken down in autumn leaves, the yellow of carotenoids becomes more visible. During the development of petals, concentrations of carotenoids increase more than 100-fold (Moehs et al. 2001). Carotenoids are chemically diverse and are produced in plastids.

Anthocyanins are flavonoids found in all tissues of higher plants. They provide red, purple, blue, and black colors to reproductive tissues. They are water soluble, odorless, and nearly flavorless, and are found most commonly in vacuoles located in, or immediately below, the epidermis. Anthocyanins are chemically diverse but absorb mostly green wavelengths. By absorbing blue, green, and UV wavelengths, they protect plants from photoinhibition, damage to the photosystem that reduces the efficiency of photosynthesis. Anthocyanins are often found in highest concentrations in flowers and fruits and were preferred by birds in artificial diets (Catoni et al. 2008). Anthocyanins have antifungal properties and may also signal to herbivores that a particular tissue is well defended (Schaefer and Ruxton 2011). A few restricted plant taxa have betalains instead of anthocyanins, and these provide roughly the same colors. Other flavonoids, the flavones and flavonols found in flowers, absorb light at shorter wavelengths and are invisible to humans. Insects such as bees can see these pigments, which form patterns of stripes, spots, and concentric circles called nectar guides (Lunau 1992). These pigments are also found in leaves and are believed to protect cells from UV-B, mediate interactions between legumes and nitrogen-fixing symbionts, and perform a variety of other useful functions.

4.4.2   Chemical cues used by animals in olfaction

Animals use volatile cues to locate plants of interest when they are searching for food, shelter, or sites where they are likely to find mates. The process of animal olfaction is complex and messy. There are several thousand plant volatiles that have been identified and this list is growing quickly (Knudsen et al. 2006). Volatiles emitted by plants are typically lipophilic compounds with high vapor pressures that can cross membranes easily (Pichersky et al. 2006). They are chemically diverse although the most common groups are terpenoids, phenylpropanoids, benzoids, and fatty acid and amino acid derivatives (Dudareva et al. 2004). The enzymes involved in producing these volatiles are themselves diverse and they often produce several different major products plus a host of minor "derailment" products; in addition, these enzymes have the unusual proclivity to act on multiple substrates (Pichersky et al. 2006). Volatiles may be produced by all plant organs—in flowers synthesis usually occurs in epidermal cells and in vegetative organs, and volatiles are often synthesized in glandular trichomes. They are sometimes produced in internal structures and stored in vacuoles, specialized cells, or ducts. Young leaves tend to have higher rates of synthesis than older ones; higher rates of synthesis are also observed when flowers are ready for pollination and fruits are ripe (Pichersky et al. 2006).

Plant volatiles are perceived by a diversity of odor receptors in animals, unlike the conserved nature of light receptors. Mammals have more than 1,000 odor receptor genes, honeybees have more than 130, and *Drosophila* have 60 (Schaefer and Ruxton 2011). Odor receptors in different animal taxa are not homologous and have little in common; receptors are typically located in insect antennas, unlike their location in the nasal cavities in most mammals (von Frisch 1919, Firestein 2001). Perception of odor cues in insects is mediated by olfactory receptor neurons that convert the chemical signals into electrical signals that are integrated and interpreted by the central nervous system (Jefferis 2005).

The nature of plant volatiles and their associated animal receptors have consequences for the transfer of information. Many receptors are extremely specific, making it likely that some plant volatiles are not perceived widely. An important consequence of this specificity is the possibility that volatile communication may provide "private channels" that are not readily available to other organisms, although this notion remains controversial (Raguso 2008). Floral volatiles allow insects to discriminate among individual flowers of a single species (Dudareva et al. 2006). Volatiles induced by herbivory are sometimes specific to the herbivore and the host plant; predators and parasites of these herbivores can discriminate among these slight variations in some instances (Takabayashi et al. 1995, De Moraes et al. 1998). Volatiles also have the potential to provide information about the time since an event took place (Schaefer and Ruxton 2011). Volatile chemicals decay at predictable rates, unlike light cues. Volatiles provide little directional information that animals can use to identify the source of cues; animals must track gradients in concentrations of the compounds to attempt to gain this information. As a result, the distances over which these gradients can be recognized may constrain the signaling system. Insect larvae have been reported to be attracted over distances ranging from 0.5–4 cm, while adults are more sensitive (Visser 1986). Mosquitoes follow odor plumes to find hosts 100 m away, and moths respond to the odors of mates at several km.

Plant volatiles usually appear in complex mixtures, often comprising several hundred compounds emitted simultaneously (Bruce et al. 2005). It remains controversial how often animals use taxonomically specific volatiles vs. ratios of volatile blends to identify host plants, although evidence suggests that the later model is more common (Bruce and Pickett 2011). The job of recognizing host plants is complicated because many of the compounds that are used as cues are ubiquitous, produced by many different species. The insects must recognize the correct blend against a background of similar compounds that are emitted by nonhosts. Honeybees are able to discriminate among snapdragon cultivars emitting the same volatiles at slightly different levels (Wright et al. 2005). Black bean aphids were repelled by 10 different volatiles emitted individually by their host but were attracted by the combination of these same compounds (Webster et al. 2010). When host recognition of blends occurs rather than recognition of specific, unique chemicals, this suggests that processing by the central nervous system plays a large role (Bruce et al. 2005). In some instances, insects can use volatile cues to recognize not only the identity of host plants but also the nutritional condition of the host and the presence of other insects on it (Bruce and Pickett 2011).

Plant volatiles have many functions in relationship to animals. They defend plants against herbivores (and pathogens), either by directly repelling the invaders (De Moraes et al. 2001, Vancanneyt et al. 2001), or indirectly by attracting predators and parasites of the plant attackers (Dicke et al. 1990, Turlings et al. 1990). They attract pollinators to fertilize seeds with out-crossed pollen and frugivores to disseminate seeds. In addition to their roles as cues for animals, some of these same volatiles have been found to protect plants from abiotic stresses (Vickers et al. 2009).

4.4.3 Chemical cues used by animals in gustation

We know far less about chemoreception and plant cues that animals taste than we do about olfaction. Unlike visual and olfactory cues that function at some distance, gustatory cues require that the animal actually interact with the plant. There are probably as many different chemicals that animals can taste as there are volatiles to smell although the number of taste receptors is quite limited. Humans and probably most mammals have five taste receptors corresponding to sweet, bitter, sour, salty, and umami (Lindemann 2001, Yarmolinsky et al. 2009). Sweet and umami (meaty or savory) are stimulants and generally promote consumption of sugar and protein whereas bitter and sour are deterrents and generally promote rejection of toxins. Salty can do either depending upon the concentration of the salt and the physiological state of the animal.

Taste receptors are membrane proteins located in the mouths of most animals although they may also be found on the legs and wings of insects. Umami and sweet are sensed by G-protein-coupled receptors that bind to a diversity of chemicals. For example, these receptors recognize simple sugars, artificial sweeteners, D-amino acids, and proteins all as sweet. Although other mammals such as mice taste umami and sweet as we do, a slightly different group of chemicals are recognized by their receptors. Mammals have on the order of 10–40 different receptors that recognize a diversity of different chemicals as bitter. The sour and salt receptors are poorly understood. Fruit flies have gustatory systems that are not homologous to those of mammals. Nonetheless, there are many similarities. Fruit flies respond to chemicals that we would call sweet, bitter, and salty. They also respond to water and CO_2.

Animals make sophisticated decisions about what to eat and what to avoid. Some of these decisions reflect innate preferences and aversions, others reflect learning. In addition, there are physiological feedbacks that influence choices that occur independently of plant cues. For example, ruminants probably don't taste the nutrients in their food but adjust their diets based on postingestive feedback to acquire the nutrients that they require (Provenza 1995). Foraging decisions made by animals integrate diverse information from visual, olfactory, and gustatory cues along with nutritional feedbacks.

The diversity of cues considered in this chapter makes it clear that plants have the ability to sense many environmental cues and make use of them to inform their plastic responses to heterogeneous conditions. Since we view the world based on our own perceptive abilities, it may be useful to compare the cues that we sense to those that plants sense. The overlap is remarkable, given how different we are from plants and how differently we acquire energy. We are more dependent on visual acuity, which is poorly developed for plants, and plants are far more dependent on chemical cues for which we have limited abilities of detection. Nonetheless, both plants and animals are affected by many of the same cues. Some of this similarity represents shared ancestry, since cellular processes including perception are ancient and highly conserved. Some of the similarity probably represents more recent adaptation since selection should favor both plants and animals that are perceptive and eavesdrop on informative cues and signals used by other organisms. As communication between plants and animals evolves, emitting cues that the receiver can already perceive will be more effective and may often be co-opted for new functions. In this context, it is not surprising that the same conserved cues are used repeatedly by many diverse organisms.

5 Plant Responses to Cues about Resources

5.1 General characteristics of plant responses

Plants both produce and respond to many cues. These cues provide the plant with information about resources that occur heterogeneously throughout the landscape. In some instances we understand the basic nature of those cues and the receptor systems that plants use to perceive them; these were outlined in the previous chapters. In this chapter we will consider the morphological, chemical, and behavioral phenotypic responses that plants exhibit when faced with cues indicating environmental heterogeneity in the levels of potential resources. Research indicates that plants selectively place their semiautonomous units (such as root, shoot, and stem modules) to respond to local conditions above and below ground, ranging from light availability to distribution of nutrients in the soil. The types of behaviors plants exhibit in response to cues are complex and varied. Because plants are faced with multiple cues, they even demonstrate the ability to make decisions about which cue to respond to when a trade-off is necessary.

Environmental conditions that are important to plants vary over space and time. Plants respond to cues that reliably indicate current or future conditions to adjust their phenotypes to match these variable environments. Phenotypic adjustments may take the form of plant movements, physiological acclimation, growth of new tissues and organs, or shedding of existing tissues and organs. These responses may be considered behaviors if they occur in response to a stimulus, are reversible, and occur rapidly within the lifespan of an individual (Silvertown and Gordon 1989, Karban 2008) (see chapter 1).

Morphological responses of plants are possible in many instances because plants are made up of repeated semiautonomous modules (White 1979, Silvertown and Gordon 1989, Herrera 2009). These modules are produced by reiterated meristems that can give rise to multiple organs of undetermined characteristics that can vary in type, size, shape, number, and function. As a result, plants are sometimes able to quickly and radically transform in response to different cues. Since plants tend to be less mobile than animals, a plastic morphology allows some of the flexibility that is lost by being rooted in one place. In addition to growing in one direction or another, plants also adjust their morphologies by shedding or abscising tissues and organs that are not as productive or valuable.

One consequence of the modular construction of autonomous building blocks is that plants are less well integrated than are most animals (Hutchings and de Kroon 1994, de Kroon et al. 2005). Different plant modules experience different conditions (light levels, herbivory, etc.) simultaneously. Plants often respond to fine-scale environmental heterogeneity with localized adjustments. In a classic example of this phenomenon, Malcolm Drew (1975) grew barley in soils with heterogeneous distributions of phosphate. Those root modules that were in contact with higher concentrations of phosphate grew more and longer lateral rootlets than other root modules of the same individual (fig. 5.1). This localized response indicated considerable independence of different modules that made up a single root. Similar results have been found for localized responses of shoots to heterogeneity in light availability (e.g., Waite 1994). Localized responses of roots were stronger when the whole plant was grown in a phosphate-poor environment. This influence of the state of the entire plant suggests some degree of integration of information among the semiautonomous units. The extent of systemic integration and the physiological mechanisms responsible are poorly understood and will repay further study.

A consequence of localized responses to environmental heterogeneity is an increase in the phenotypic variation in individual plants. Plant responses to variation are often depicted as norms of reactions for the entire individual (e.g., Via 1987, Harvell 1990). These norms of reaction models show changes in phenotype means in two environments but quite often do

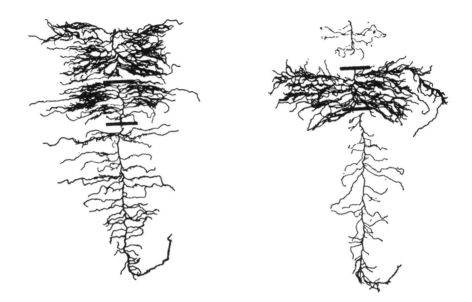

FIGURE 5.1 A root profile of barley plants grown in sand that either had high levels of nutrients throughout (left) or had high levels of phosphate located only in the middle zone of sand between the two horizontal lines (right). Roots proliferated selectively in the zone with high levels of phosphate. (From Drew 1975).

not include variation around the means within an individual. This representation of induced responses and other forms of plasticity implicitly assumes systemic integration throughout the individual rather than localized responses among plant organs to fine-scale environmental heterogeneity. Localized plant responses have the effect of increasing phenotypic variation among the various tissues and organs of the responding individual (fig. 5.2) (Stout et al. 1996, de Kroon et al. 2005). This increased plant variation may be difficult for pathogens and herbivores to exploit in both ecological and evolutionary time frames (Adler and Karban 1994, Karban, Agrawal, and Mangel 1997). Increased plant variation may itself allow plants to appear as moving targets in time and space, challenging the abilities of pathogens and herbivores to accommodate.

5.2 Plants forage for resources

Plants place and remove leaves, roots, and reproductive structures nonrandomly within their heterogeneous environments, and this placement allows them to effectively forage for light, water, mineral nutrients, and

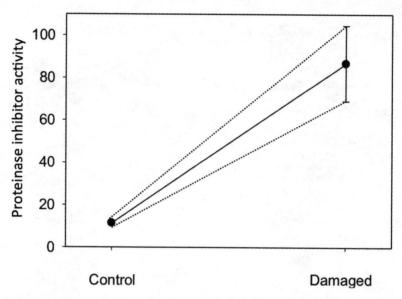

FIGURE 5.2 Proteinase inhibitor activity in tomato leaflets that were adjacent to experimental damage caused by *Helicoverpa zea* caterpillar feeding or adjacent to an undamaged control. Activity is expressed as mean % inhibition relative to a control with chymotrypsin alone. Error bars show standard errors and were present for both control and damage treatments. Following damage, the mean increased 8-fold and the standard error increased 60-fold. Data from Stout et al. (1996).

reproductive services (Grime 1979). The placement or abscission of leaves and roots is not strictly reversible since plants cannot completely undo their investments in these structures once they have been produced. However, they can certainly change the direction of allocation decisions, and they may recover some of the nutrients that were invested in these organs when they are no longer valuable to the plant before these tissues are shed.

5.2.1 Foraging for light (shade avoidance)

The light reaching a plant canopy is highly variable over short distances and short time spans. Levels of radiation reaching the understory may be 100 times as great in light gaps as under dense canopy vegetation (Smith 1982). Leaves at the top of the canopy receive far more light than those on shaded branches lower in the canopy of the same tree. Light quality also changes from a red:far-red ratio of 1.15 in full sunlight to 0.05 under dense vegetation (Hutchings and de Kroon 1994). Plants sense the quality and quantity of incoming light using phytochrome receptors (see section 2.2.2). Phytochrome

TABLE 5.1 Shade avoidance responses to low light conditions

Developmental stage	Response	Consequences
seeds	no germination	seed dormancy
seedling	hypocotyl elongation	emergence from soil
	hooked hypocotyl	protection of the meristem
vegetative	elongated internodes	rapid vertical growth
	few leaves	rapid vertical growth
	upward leaf movement	intercept more light
reproductive	early flowering	earlier production of progeny
	early seed maturation	progeny to more favorable environment

acts as a local signal within individual cells and regulates the expression of many genes (see section 4.2). Other more mobile signals, most notably ethylene, are stimulated by shading and regulate gene expression systemically in other parts of the plant.

Low levels of light or light with a low red:far-red ratio initiate the shade avoidance responses in many plants (table 5.1). It should be kept in mind that these responses are widespread but not universal and depend on the developmental stage of the plant. "Shade plants" that are adapted to low-light conditions are far less responsive to light cues than are "sun plants" that are adapted to high-light conditions. Seeds that experience a low red:far-red ratio, characteristic of shady environments, fail to germinate until light cues indicate a more favorable environment (Vazquez-Yanes and Smith 1982). Hypocotyls of shaded seedlings elongate rapidly and leaves do not expand as seedlings attempt to grow out of the dark soil and into a lighted environment. During this time, the hypocotyl maintains a hooked morphology that helps the seedling push out of the soil while protecting the apical meristem. Older plants that experience shading form long internodes and relatively few leaves. This causes shaded plants to grow taller and less bushy. Leaves of shaded rosette plants bend upward to intercept more light. In contrast, plants that experience favorable light conditions reduce the length of vertical internodes and increase lateral branch formation. If these shade-avoidance responses are successful, the plant will have grown above its competitors or other impediments that can block its access to light. If these elongation responses fail to get it into unattenuated daylight, plants will accelerate flowering and production of seeds to produce progeny that can escape the unfavorable conditions (Halliday et al. 1994, Cerdan and Chory 2003). Early flowering caused by low red:far-red is associated with reduced seed set, fruit development, and seed germination, but presumably reproducing increases plant fitness more than would continued shading.

More and larger buds develop on branches in sunny patches than on branches in shady patches, resulting in asymmetric growth. Shade avoidance responses result in plants filling light gaps where there are fewer neighbors competing for light (fig. 5.3) (Harper 1985, Novoplansky et al. 1990, Young and Hubbell 1991). Plants also shed those organs that are not receiving light and not actively growing, which accentuates their bias towards filling of canopy space where light is most available. Plants respond similarly whether reacting to shade cast by their own leaves or those of another individual.

Several features of this light foraging are particularly noteworthy. First, plants appear to differentiate between shade cast by an inanimate object in their environment and a photosynthetically active competitor. *Portulaca oleracea* seedlings grew away from any object that cast shade, but mature plants responded more strongly to shade cast by photosynthetically active plant tissue than inanimate objects (Novoplansky et al. 1990). Seedlings of *P. oleracea* grew preferentially towards light with a higher red:far-red ratio rather than towards a high photon flux density (Novoplansky 1991).

Second, plants respond in such a way as to appear to anticipate future competitors. Seedlings of *Datura ferox* and *Sinapis alba* responded to neighboring plants well before actually being shaded (Ballare et al. 1990). They responded to the far-red light that was being scattered by surrounding foliage before levels of light intensity were reduced and gained information about the relative positions of neighbors at a very early stage in canopy development (Ballare et al. 1990). In contrast to red light depletion, blue light depletion does not occur prior to shading and indicates actual attenuation of light rather than future shading (Pierik, Whitelam, et al. 2004). These two systems allow leaves to both anticipate future competitors and to evaluate current conditions.

Third, the fitness consequences of the shade avoidance syndrome have been particularly well studied (see section 9.2.2). Plants with the appropriate phenotype for their environment (high or low density of competitors) were more fit in those environments (Dudley and Schmitt 1996). However,

FIGURE 5.3 Branching by a pair of *Tilia* trees growing close together. Branches grow away from competitors and those growing towards competitors are shaded and subsequently shed. Redrawn from a photograph by Michelle Jones in Harper (1985).

plants that invested in greater stem growth in response to competitors that shaded them invested less in root development; as a result, the benefits and costs of the shade avoidance response also depended on the plant's access to moisture (Huber et al. 2004).

The amount of light reaching a leaf varies greatly over time as well as space as the sun passes through the sky and clouds block incoming radiation. In situations where plants experience ephemeral pulses of resources they respond physiologically rather than with morphological plasticity. For example, trees in temperate and tropical forests receive 40–90% of their photon flux density from highly ephemeral sun flecks, which are ephemeral light beams that create patches of much higher irradiance than the leaf usually experiences (Chazdon and Pearcy 1991). In these cases photon flux density changes by more than 100-fold within a few seconds. Physiological plasticity to sun flecks is more rapid and less permanent than morphological adjustments; physiological plasticity is probably more common for plants that are adapted to environments that are typically lower in light while morphological plasticity is more common for plants adapted to high-light environments, although this hypothesis has not been rigorously tested (Hutchings and de Kroon 1994).

Plants respond to increased light with two kinds of physiological responses that occur over two different time scales. Light activates rubisco and other enzymes that are required for photosynthesis over time scales of seconds to minutes (Pearcy et al. 1994). Light also causes stomata to open so that the gas exchange required for photosynthesis becomes possible. By building up pools of intermediates immediately after illumination that can be used later, leaves continue to fix CO_2 after a sunfleck has passed. Plants also adjust concentrations of enzymes and leaf anatomy over time scales of days to weeks (Pearcy and Sims 1994). Acclimation to prevailing light levels increases photosynthetic rates at any given photon flux density. In addition, new leaves that are grown under conditions of more light are thicker, which allows them to have increased photosynthetic capacity.

Charles Darwin noticed that all plant organs undergo subtle movements around their axes of elongation, which he termed circumnutation (Darwin 1880). By modifying this behavior, plants can direct movements in response to light, a phenomenon that we now call phototropism. This allows them to forage for light more efficiently than individuals that were prevented from such directed movements. Darwin identified the tip of the shoot as the organ that was sensitive to light. This observation led to the discovery of the first plant hormone, auxin, which is produced in the shoot tip but moves down the stems and controls growth and development (Went 1926). Auxin accumulates on the dark side of the shoot, causing cell expansion so that the shoot tip bends towards the light source. Even before Darwin, Julius von Sachs realized in 1864 that plants were most responsive to blue wavelengths (Christie and Murphy 2013). Blue light is sensed by a family of receptors that are protein kinases located in cell membranes (Hohm et al. 2013). These protein kinases trigger the auxin gradient that leads to asymmetric growth. Remarkably, plants can detect and respond to subtle gradients in light intensity even when incoming light intensity spans at least three orders of magnitude.

Other plant organs exhibit related behavioral responses to light. Responses to blue light affect stomatal opening and intracellular movement of chloroplasts (Hohm et al. 2013). Flowers and leaves bend towards the sun on a daily basis, a process that is independent of growth and is referred to as heliotropism (Christie and Murphy 2013). Other plant organs such as roots and shoot tips of climbing plants grow away from light, directing them towards soil and structural supports, respectively.

5.2.2   Foraging for nutrients

The distribution of nutrients in the soil is also extremely patchy. The underlying heterogeneity is magnified because roots deplete nutrients in their vicinity, producing zones of influence where resource levels are low (Casper et al. 2003). Roots are modular units, like shoots, that can respond locally to small-scale heterogeneity. Plants selectively place lateral roots into areas that are relatively richer in nutrients (Drew et al. 1973, Drew 1975, Nibau et al. 2008) (see fig. 5.1). Plants also leave nutrient-rich patches less quickly than nitrient-poor patches, although the demographic responses of roots are poorly known (Gross et al. 1993, Cahill and McNickle 2011). Unlike lateral roots, which are responsive to nutrient levels, extension growth along the main root axis is often unaffected by nutrient availability. This pattern of growth suggests that roots continually search through the soil for nutrients but form more lateral roots when rich patches are found.

Root topology changes depending upon patch quality. For example, a herring bone topology (branching predominantly on the main root axis) is most effective for exploring the soil for rich patches but less effective once such a patch has been encountered (Farley and Fitter 1999). Thinner and sometimes longer roots were found when nutrients were less abundant; these explored the soil more efficiently than shorter thicker roots (Hutchings and de Kroon 1994). Differential root placement and demography caused roots to become aggregated in high-quality patches (Cahill and McNickle 2011). Conversely, roots may become segregated from one another as a result of avoidance behaviors and allelopathy. Examples of both aggregation and overdispersion of roots have been observed. For instance, soybean plants produced more roots when grown with competitors (Gersani et al. 2001) while *Ambrosia dumosa* placed roots away from conspecific competitors (Mahall and Callaway 1991).

Roots detect and avoid inanimate objects like rocks (Falik et al. 2005). Pea roots are thought to respond to their own exudates, which accumulate near impermeable obstacles. Lateral roots were shorter in the direction of the obstacle and many of them subsequently withered. Treatments that reduced the accumulation of root exudates failed to show the avoidance responses near obstacles. Similar results were also found for grasses from resource-poor habitats that encountered obstacles (Semchenko et al. 2008).

Roots encounter numerous local cues that provide potentially useful information about levels of resources that will likely be present in nearby places and times. Plants appear to have some ability to coordinate their responses by integrating various bits of environmental information. For example, pea seedlings were cultivated so that individual plants had their roots split equally and placed into two pots (Gersani and Sachs 1992). Initiation of lateral roots was determined by the relative amounts of resources in each pot, not the absolute amounts. Plants with their roots split between pots with different levels of competitors proliferated roots in those pots with fewer competitors (Gersani et al. 1998). Evidence suggests that plants use diverse cues in addition to resource depletion to perceive the presence of competitors for resources (Schenk 2006). For example, barley plants responded to volatile cues to adjust allocation to roots even when direct interactions between roots were prevented (Ninkovic 2003).

The responses of roots depended upon the identity of neighboring competitors. For example, *Ambrosia dumosa* individuals responded differently to roots of conspecifics than to roots of heterospecifics (Mahall and Callaway 1991). Plants discriminate between their own attached roots and those of other individuals, and they avoid competing with themselves by growing shorter and fewer roots or by growing away from other self roots (Falik et al. 2003, Holzapfel and Alpert 2003, Gruntman and Novoplansky 2004). Self-recognition in these cases required the roots to be physically attached; separating them resulted in a failure to recognize them as self. Some plants distinguish the roots of kin and strangers. Individuals of *Cakile edentula* that encountered roots of kin allocated fewer resources to fine roots than individuals that encountered the roots of strangers (Dudley and File 2007). The ability to distinguish kin from strangers is important because it suggests that plants could potentially behave preferentially towards kin relative to strangers, making cooperation and other social behaviors more likely to evolve (Hamilton 1964).

The mechanisms by which roots perceive and respond to nutrient patches are not as well understood as the shade avoidance responses of stems. Several of the genes that are involved in the perception of soil nutrients have been identified along with their tissue-specific locations in the roots (Hodge 2009) (see also section 2.3.1). However, the biochemical and physiological mechanisms responsible for these processes remain unknown.

It stands to reason that plants that can forage efficiently for resources will experience a fitness advantage over individuals that are less capable in this regard. Plants that were the best able to proliferate roots in response to nitrate patches were more competitive at capturing nitrogen (Hodge et al. 1999, Robinson et al. 1999). These competitive benefits were found to be highly context dependent. In addition, because these presumed adaptations occur underground and out of sight, we know very little about the fitness consequences of root foraging, far less than about foraging for light. One study that has addressed this issue compared the fitness of an *Arabidopsis* mutant that had limited ability to produce lateral rootlets in greenhouse and field experiments (Fitter et al. 2002). This mutant experienced reduced fitness in soils that were limited by phosphate, which is not mobile in soils. In contrast, the ability to produce lateral rootlets was not associated with a fitness advantage in soils with limited nitrate, which is water-soluble (mobile) and could be harvested effectively by mutants without extensive lateral rootlets. However, in more competitive situations, lateral roots may also aid in acquiring mobile resources like nitrate (Hodge 2004).

Root proliferation is a relatively slow process, requiring on the order of tens of days. Roots also respond physiologically to heterogeneous soil nutrients; these responses can be much more rapid, on the order of hours to days. Uptake of mobile ions, such as NO_3^-, and water is thought to be strongly influenced by physiological adjustments (Hodge 2004). Local experimental enrichment of soil with various nutrients resulted in rapid increases in rates of uptake of the nutrient supplied (Jackson et al. 1990, Fransen et al. 1999). When plants encounter high levels of soil nitrate, they activate nitrate transporters that allow them to more effectively acquire this nutrient (Okamoto et al. 2003, Gan et al. 2005).

Plants often rely on symbiotic mycorrhizae to assist with acquiring resources, and these fungi proliferate hyphae into resource-rich patches. Infection of roots with arbuscular mycorrhizal fungi reduced root branching and production of lateral roots and root hairs in some species (Hetrick et al. 1991, Schweiger et al. 1995). This suggests that mycorrhizal fungi may take the place of root morphological and physiological responses in these cases.

Roots release exudates in response to mineral deficiencies that alter the soil solution and allows enhanced uptake of nutrients (Metlen et al. 2009). These responses are particularly well studied for uptake of P, which tends to be bound to positively charged soil minerals (often metals) (Hinsinger 2001). In response to low P availability, plants of many families but especially legumes release biochemicals that bind (or chelate) to the metals, making the P more biologically available. In calcareous soils a similar process involving root exudates binds to the soil Ca, making P more available for plants. Plant responses to P limitation occurred rapidly over a range of hours to days and were reversed when conditions changed (Metlen et al. 2009). Secretions were localized to the precise soil locations where the root segment contacted recalcitrant P (Hoffland et al. 1992). Roots also responded with localized secretions of secondary metabolites to suboptimal soil concentrations of aluminum and iron; these responses improved conditions for acquisition of needed resources (Metlen et al. 2009). Soybean plants that were grown in a stratified medium with different concentrations of P and Al secreted the specific metabolites necessary for the acquisition of each nutrient in the appropriate region of the root (Liao et al. 2006).

5.3   Integrating resource needs

Plants require multiple resources to be successful, and they actively forage for light, water, and mineral nutrients. This raises several questions: What happens when foraging for one resource makes foraging for another less efficient? Are plants able to coordinate and prioritize resource needs without a central nervous system?

Plants allocate internal resources to achieve a functional equilibrium between the needs of roots and stems. Plants that are limited by mineral nutrients or water invest in the production of more roots, and those limited by light or carbohydrates invest in the production of more shoots and especially leaves (Brouwer 1963, Bloom et al. 1985). This pattern of balanced allocation was found for plants of all habitats, growth rates, and competitive abilities (Reynolds and D'Antonio 1996, Poorter and Nagel 2000). An increase in allocation to roots when nutrients were limiting was found to be particularly strong and highly conserved among plant species. As water became less available, plants were able to reduce allocations to leaves so that photosynthesis was not greatly affected over moderate levels (Boyer 1970).

5.3.1   Integrating information about light and nutrient acquisition

Evidence suggests that plants are able to integrate information from several different sources to make allocation decisions. For example, when *Abutilon theophrasti* individuals were grown alone, their roots were distributed without strong dependence on the distribution of soil nutrients (Cahill et al. 2010). However, when individuals were grown with neighbors, they adjusted their placement of roots and foraged less widely. With competitors, root growth was also more sensitive to the direction of nutrients and the direction of neighbors.

Although localized shoot or root modules perceive and respond to localized patches of resources at very fine scales (de Kroon et al. 2005), there is evidence that they also integrate information about the overall state of the individual in their foraging decisions (de Kroon et al. 2009). Only by reacting to cues at very fine spatial and temporal scales can plant responses be precise enough to track environmental heterogeneity. However, feedback from other modules can influence the transport and effect of localized cues, hormones, and cofactors involved in plant responses.

Although the details of such feedback systems are not understood, there is considerable evidence of their existence. Shade avoidance reactions, such as changes in internode elongation and upward leaf movement, involve localized phytochrome detection and responses that occur at very fine scales. However, these responses are affected by the overall systemic level of shading that the plant experiences. Plants that are shaded transport auxin preferentially in the outer layers of cells and less in the central core of the shoot (Morelli and Ruberti 2002). This change in the location of auxin transport results in increased responsiveness in stems but reduced responsiveness in leaves. Systemic levels of sugars also influence localized responses to shade. For example, when systemic carbohydrate concentrations became low, plants became less responsive to shade (Kozuka et al. 2005, Smith and Stitt 2007). An adaptive explanation for this lack of responsiveness is that elongation would have further stressed the plant and would have required carbohydrates that the plant did not have.

Localized responses of roots are also affected by the conditions experienced by the whole plant. The localized responses to soil nitrate were greatest when the overall nitrogen status of barley plants and tree seedlings was low (Drew et al. 1973, Friend et al. 1990). Specific responses of *Arabidopsis* roots have also been found to be very different depending upon the overall resource status of the plant. Rates of proliferation of lateral roots were increased by contact with localized nitrate patches (Zhang and Forde 1998), although high rates of nitrogen supply to the whole plant suppressed lateral root development (Zhang et al. 1999). Several feedback mechanisms operating at the level of the whole plant are thought to strongly influence localized nitrate uptake (Forde 2002). These include nitrate transporters, as well as auxin and sugars supplied from the shoots. Depriving roots of nitrate limited the responsiveness of the shoots to cues that would cause leaf expansion under more favorable conditions (McDonald and Davies 1996).

Experiments with clonal plants also indicate that individual ramets integrate information about local conditions with information about the state of the larger collection of connected ramets. *Solidago canadensis* ramets exhibited patterns of growth and reproduction that depended on the species identity of their neighbors (Hartnett and Bazzaz 1985). When interconnected ramets were grown with a diversity of neighbor species, the ramets all responded similarly, with the average of the responses they would be expected to show to the individual neighbor species. Connected ramets of clover were placed in environments that differed in the resources that they provided (Stuefer et al. 1996). Individual ramets specialized in acquiring the resource that was locally most abundant and "traded" with other ramets for resources that were less locally abundant, rather than each ramet attempting to acquire them all. These examples indicate that plants are able to integrate numerous sources of information about the overall status of the individual along with fine-scale information about the status of localized sections of shoots or roots.

5.3.2   Integrating information about resource acquisition and defense

Plants not only need to balance the various demands of acquiring multiple resources and distributing them to accommodate local and systemic needs. Acquiring resources also affects other important processes such as defense and reproduction. Plant responses to herbivore attacks and to reproductive needs are considered in the following two chapters. In some instances, trade-offs have been observed between foraging for resources and defending against attackers although these trade-offs are not universal. Such trade-offs are considered in this chapter because, in many instances, resource acquisition has been found to take precedence over defense.

Individuals of *Chenopodium album* that were grown under light conditions indicative of competitors were better hosts for caterpillars (Kurishige and Agrawal 2005). The converse was not found; plant responses to herbivory did not limit subsequent responses to shading. Similarly, shade-intolerant tobacco and tomato responded to far-red radiation, an indication of competitors, by allocating resources to shade avoidance responses (Izaguirre et al. 2006). These responses were associated with downregulation of chemicals thought to provide defense against herbivores and with increased growth rates for a specialist caterpillar. *Arabidopsis* also responded to far-red radiation by reducing defenses against herbivores (Moreno et al. 2009). In this case, plants exposed to far-red were less sensitive to jasmonate signals that activate many defensive responses. Similar reductions in responsiveness to SA have also been observed for plants experiencing shading (De Wit et al. 2013). It is possible that the phytochromes that recognize altered light quality associated with competitors also affect regulatory action of the hormones—JA, ethylene, and SA—that control induced defenses against herbivores and pathogens (Genoud et al. 2002, Pierik, Cuppens, et al. 2004, Griebel and Zeier 2008). The interactions between phytochromes and JA are particularly interesting since JA has long been recognized as an inhibitor of cell division and elongation (Koda et al. 1992, Yan et al. 2007, Zhang and Turner 2008).

These results suggest that at least some plants prioritize avoiding shade over defending against herbivores and pathogens. This prioritization appears to be mediated by phytochrome perception of light quality that reduces the plant's sensitivity to JA and SA induced by herbivore damage and pathogen attack (Ballare 2009, 2011, De Wit et al. 2013). Less is known about the consequences of responding to attack on resource acquisition.

In general, acquiring resources is crucially important for plants, and they have evolved sophisticated systems for evaluating resource quality and responding appropriately. They appear to forage efficiently for light and nutrients in a variety of environments and situations. However, the extent to which plants integrate diverse sources of information and the processes by which they prioritize different needs are research questions that are poorly explored.

6

Plant Responses to Herbivory

6.1 Induced responses as plant defenses

A key evolutionary challenge for plants is attack by herbivores. Plants defend themselves by changing their phenotypes when they sense cues of impending risk, such as previous damage. Some of the cues that plants use to sense herbivores were considered in section 4.3, including changes in membrane potentials and ion fluxes, plant hormone titers such as JA, and volatiles. Plant responses to herbivores may involve morphological characteristics including leaf toughness, density of spines, or the placement of vulnerable meristems; chemical characteristics including primary metabolites, toxins, or compounds that interfere with digestion; and phenological characteristics such as the timing of leaf expansion and abscission, or allocations to reproduction (Karban and Baldwin 1997). Induced responses have been described from virtually all plant species that have been examined, and they involve a wide diversity of plant traits. Many of the recent reviews of this field have concentrated on biochemical mechanisms of induced responses to herbivory (Howe and Jander 2008, Mithofer and Boland 2012). Thus, this subject will not be considered here. The various responses to herbivory are thought to provide induced resistance if they cause herbivores to be less attracted to induced tissue or if they cause herbivores to perform less favorably following consumption of these tissues (Karban and Myers 1989). Not all of the plant responses to damage are harmful to herbivores; some result in herbivores experiencing improved performance on damaged plants, a phenomenon called induced susceptibility (Karban and Baldwin 1997, Nykanen and Koricheva 2004). Other induced responses have no particular consequence to herbivores but involve healing wounds, enhancing resource capture, and so on. Since induced resistance is evaluated from the herbivore's point of view, induced responses that provide resistance need not necessarily benefit the plant. Induced responses that allow plants to experience higher fitness in environments with high risk of herbivory are considered induced defenses (Karban and Myers 1989).

As was true for resources, risk of herbivory is often heterogeneous over time and space—herbivores are abundant at some times and absent at others; some locations have many, others few. A plant that consistently faces either very high or very low levels of herbivory is expected to evolve constitutive defenses that are permanently expressed and match that level of risk (Adler and Karban 1994, Karban and Nagasaka 2004, Ito and Sakai 2009). Induced defenses are hypothesized to allow plants to save resources or other costs when those defenses are not needed and to rapidly deploy them when they are. This reasoning echoes earlier arguments about the advantages of plasticity over fixed traits in environments that are variable and unpredictable (Darwin 1881, Bradshaw 1965). In a letter to Karl Semper, Darwin wrote, "I speculate whether a species very liable to repeated and great changes of conditions might not assume a fluctuating condition ready to be adapted to either condition."

For plant responses to herbivores to be beneficial, plants must be able to respond to reliable cues of impending risk and respond appropriately. In this chapter we will consider some of the sources of information about risk of herbivory that plants can obtain and the evidence that responding to these cues increases fitness. We will also examine some of the plant cues that herbivores can detect.

6.1.1 Plant defenses are context-dependent

There are many caveats and conditions that must be satisfied before induced responses will actually provide defense for plants; we will examine them in turn over the next sections. First, deploying effective defenses is complicated because many of the plant traits that ultimately provide defense are highly context dependent. For example, the induced proteinase inhibitors

(PIs) in tomato foliage can be difficult for herbivores if protein is limited but have little effect is protein is abundant. Furthermore, interactions between PIs, herbivore gut pH, and oxidative enzymes can completely deactivate PIs (Duffey and Felton 1989, Felton et al. 1992). The take-home message is that plant traits are not inherently defensive but may provide defense under a limited set of chemical, environmental, and biological conditions. Traits that provide defense against one herbivore may be ineffective against another at most concentrations and may even be sequestered by some specialist herbivores to their own advantage (Lankau 2007, Ali and Agrawal 2012).

6.1.2   Cues may provide reliable information

Second, induced responses will be effective only if plants respond to cues that reliably predict future risk of herbivory. If individuals respond to cues that are unreliable and don't indicate the true risk of attack, then those individuals will experience the costs of constitutive defenses without the benefits. Conversely, if plants fail to detect or respond to cues that predict actual risk of attack, then they will express a relatively undefended phenotype, which herbivores can presumably exploit.

Surprisingly little is known about the reliability of information that plants use to induce responses to herbivory. One widespread cue that induces plant responses against herbivores is previous damage. If a plant is getting damaged now, the same herbivore is likely to damage it in the near future. Early season damage by a multivoltine leaf-mining caterpillar was a good predictor (explaining 32% of the variance) of the risk of damage later in the season for a population of wild cotton that showed evidence of induced resistance (Karban and Adler 1996). Early season damage by leaf-miners was a better predictor of later season damage by these specific herbivores than was the average of damage levels caused by all herbivores. Similarly, if one part of a plant is being attacked, then other parts of that same individual may be at higher risk. In this wild cotton example, information from the entire plant was more informative at predicting damage to an individual shoot than was damage solely from that shoot, and there was no indication that the value of the information decayed over the growing season. In a second example involving a tropical shrub, rates of early season attacks by gallmakers were strong predictors of later season galling ($R^2 = 0.51$–0.64) and early season chewing damage was a strong predictor of later season chewing ($R^2 = 0.26$–0.34) (Cornelissen et al. 2011). These are the only attempts that I am aware of to test the assumption that past herbivory is a reliable predictor of future risk. For other herbivores that exhibit limited dispersal, eggs are likely a good predictor of impending damage by feeding larvae (e.g., Beyaert et al. 2012) and early season damage is likely to be a good predictor of later season damage (e.g., Dalin and Bjorkman 2003), although these assumptions have rarely been tested.

In some unusual instances, current herbivory might be a reliable cue that risk of future herbivory is reduced. This scenario is possible when herbivores are migratory and plants experience one concentrated bout of herbivory at some point in the season followed by a much reduced risk of repeated attacks. This appears to be the case for several plants that are heavily grazed by migratory mammalian herbivores. Some populations of *Ipomopsis aggregata* and *Sanicula arctopoides* are grazed heavily by migratory elk and deer at only one point early in the season (Paige 1992, Lowenberg 1994). Similarly, some populations of *Gentianella campestris* are grazed by horses early in the season (Lennartsson et al. 1998). Plants in these populations refrain from flowering until after they experience herbivory. In these cases, tissue loss caused by herbivory is probably a reliable cue that risk of future attack is diminished. In all three of these examples, grazed individuals produced more fruits and seeds than individuals that had not been grazed, presumably because ungrazed individuals were saving resources until after they received a reliable cue that the risk of grazing had past. In contrast, other populations of *I. aggregata* that experienced repeated herbivory throughout the season did not respond to the initial damage by initiating reproduction (Maschinski and Whitham 1989, Paige 1992). In these populations, herbivory was not a reliable cue that future risk was diminished.

Induced responses may be favored by selection if past or current herbivory reliably predicts future risks, at least much of the time. Although previous herbivory events are the most obvious cues, plants may respond to any cue that they can detect that reliably predicts future conditions (Karban et al. 1999). As previously mentioned, eggs of herbivores may be a very reliable cue that the plant is likely to experience feeding in the near future, and plants respond to oviposition cues by mounting defenses (Kim et al. 2011, Beyaert et al. 2012). *Brassica nigra* plants responded to incidental cues of moving snails (mucus) and became less palatable to snails before they were actually attacked; presumably mucus was a reliable predictor of increased risk (Orrock 2013).

Plants may also respond to diverse cues that are less directly connected to herbivores as long as they are reliable. For example, plants may use the presence of carnivores to indicate a lowered risk of herbivory. Individuals of *Piper cenocladum* produced fewer chemical defenses when protective ants were present than when they were absent (Dyer et al. 2001). Plants may also use cues about their abiotic environment to predict risk of herbivory and to adjust their defenses accordingly. For example, risk of herbivory is predictably higher for *Cardamine cordifolia* in sunny microsites (Louda and Rodman 1983a). Moving plants to sunny microsites caused them to increase concentrations of chemicals that provided defense against many of these herbivores (Louda and Rodman 1983a, b, 1996). In both of these examples, it was not clear whether plants were responding to environmental information (cues) or responding directly to favorable environmental conditions.

6.1.3 Induced responses may increase plant fitness

Third, induced responses will be effective only if they deter herbivory, reduce plant damage, or make plants more tolerant of the damage that they receive. Cues may be reliable and responses may affect herbivores, but they must also consistently increase plant fitness to be favored by selection. For example, plant traits that reduce nutritional quality for herbivores may cause those herbivores to consume more plant tissue, but this scenario is unlikely to benefit the plant or be favored by selection. The empirical evidence that plant responses to cues increase plant fitness is surprisingly limited (Agrawal 1998a, Baldwin 1998, Karban 2011).

Plants can benefit from their induced responses if herbivores are sensitive to plant quality and avoid plants that are induced. Herbivore behavior is critical for effective plant defenses, although this step in the process has received relatively little attention (Adler and Grunbaum 1999). It has long been appreciated that some herbivores avoid local sites that have been previously damaged to feed preferentially on tissues that have not been attacked (Edwards and Wratten 1983). In some instances, herbivores respond to volatiles emitted by damaged plants and avoid those plants without having to sample them (De Moraes et al. 2001, Horiuchi et al. 2003).

Behavioral responses of herbivores to induced variation in plant quality and the consequences of those behaviors on plant fitness are surprisingly poorly studied. Studies have found that the scale at which herbivores respond to cues of plant quality can be critically important in terms of plant fitness. For example, some herbivores respond to variation within individual leaves (Whitham 1983, Orians et al. 2002, Shelton 2005). Experiencing a different spatial distribution of herbivory does not necessarily affect plant fitness. However, several studies have found that the distribution of damage can have as large an impact on plant fitness as the total amount of damage that herbivores inflict (Lowman 1982, Watson and Casper 1984, Marquis

1992, Mauricio et al. 1993, Lehtila 1996). Herbivores also sometimes avoid plants at a scale of individuals or neighborhoods; this behavior has the potential to make induced responses particularly valuable to plants (Barbosa et al. 2009).

6.1.4 Plant responses occur at different spatial scales

Many induced responses to herbivory are systemic; damaging one leaf makes other leaves on the plant more resistant (Karban and Baldwin 1997). The first observations and reports of induced resistance involved systemic responses to damage in tomato (Green and Ryan 1972, Pearce et al. 1991), birch trees (Haukioja and Hakala 1975, Haukioja and Neuvonen 1985), and cotton (Karban and Carey 1984). These early reports established a precedent and expectation for this field. Since then, many induced plants responses have been found that are localized, and resistance does not spread equally throughout the plant. Even for these classic systems, more recent work has revealed stronger responses in some regions of the plant, particularly those closely connected to the site of damage (tomato—Orians et al. 2000; birch trees—Tuomi et al. 1988; cotton—Karban and Niiho 1995). The emerging generalization is that most plants are sectored to some extent. Nutrients, secondary chemicals, and hormones are not readily exchanged among spatially distant plant tissues. Exchange occurs most freely and rapidly between tissues that share vascular connections (Orians and Jones 2001, Schittko and Baldwin 2003). Since vascular connections between tissues are limited, these conduits constrain transfer of the defensive compounds and signals that determine induced resistance (Viswanathan and Thaler 2004, Orians 2005, Rodriguez-Saona and Thaler 2005).

Another constraint on the propagation of signals by vascular connections is that they are more effective when plants are actively transpiring, which allows xylem transport. Plants that grow in arid environments reduce transpiration during much of the time, and these species tend to be highly sectored (Waisel et al. 1972, Zanne et al. 2006, Schenk et al. 2008). Plants from arid growing regions are unlikely to be capable of systemic responses using vascular cues. They may rely on volatile communication to overcome this constraint (see section 6.2).

One important consequence of plants being highly sectored is that modular units are largely autonomous and poorly integrated. This has the effect of increasing within-plant heterogeneity, particularly after different parts of the plant have responded to different localized stimuli (see section 5.1).

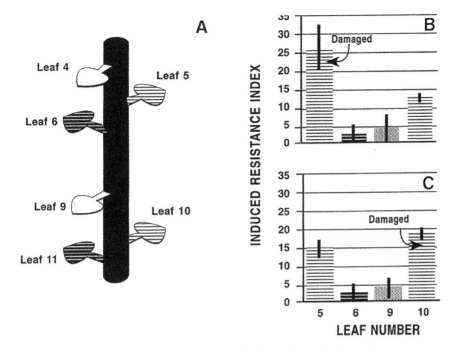

FIGURE 6.1 Vascular connections constrain communication and induced resistance in cottonwood trees. A. Every fifth leaf is connected, and the connected leaves share a symbol in the figure. B. Following experimental damage to leaf 5, those leaves that are connected (5 and 10) show induced resistance but other adjacent leaves (6 and 9) do not. C. The signal moves up or down the shoot so that when leaf 10 is experimentally damaged, connected leaves 5 and 10 show induced resistance. Figure from Karban and Baldwin (1997) drawn from data in Jones et al. (1993).

In some instances, it has been possible to map the spatial extent of phenotypic changes within individual plants. For example, responses of young tomato plants to damage by different herbivores revealed a pattern of defensive phenotypes, measured by enzyme activities, which varied greatly among leaflets and leaves (Stout et al. 1996). The mean level of activity of oxidative enzymes increased for the entire plant but responses were localized such that within-plant variation in enzyme activity increased even more than mean levels—plants became more heterogeneous (fig 5.2). Limitations in vascular connections probably contributed to this increased variability (Orians et al. 2000). For example, every fifth leaf of a cottonwood tree shares vascular connections (Jones et al. 1993). Following damage to leaf 5, induction was noted in leaf 10, which shared vascular connections with leaf 5, but not in the adjacent leaf 6, which failed to share direct vascular connections (fig. 6.1). At an even finer scale, the side of the leaf closest to the direct vascular connection responded more than the side farther away. The

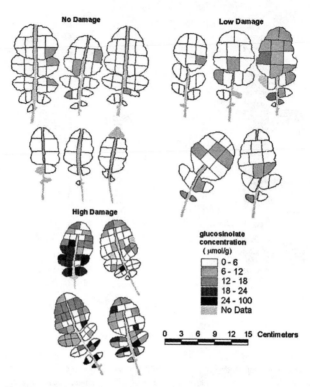

No Damage

Low Damage

High Damage

glucosinolate
concentration
(µmol/g)

0 - 6
6 - 12
12 - 18
18 - 24
24 - 100
No Data

0   3   6   9   12   15   Centimeters

FIGURE 6.2 Heterogeneity in the glucosinolate concentrations of young radish leaves on plants that received no damage, low levels of damage, and high levels of damage from *Pieris rapae* caterpillars. Induction from herbivore damage makes subsequent leaves more variable in their defensive profiles. Figure from Shelton (2005).

glucosinolate responses of wild radish to damage showed heterogeneity at very fine spatial scales (fig. 6.2). Caterpillar feeding on one leaf increased the mean concentration of glucosinolates for the entire plant but differences among regions within leaves and among different leaves were far greater than between-plant differences. No spatial autocorrelations were found in glucosinolate concentrations among leaf tissues as close as 1–2 cm apart (Shelton 2005).

6.1.5   The timing of induced responses to herbivory

The timing of induced responses influences their effectiveness. At least two aspects of timing play critical roles—the speed of induction and the timing of relaxation. All induced responses require some lag between the time when the cue is detected and the time when the defensive phenotype is expressed. This interval of relative vulnerability reduces the value of induced responses relative to constitutive defenses. The lag time before induction varies greatly among both plant signals and responses; both can range from seconds to years. Different plant signals require different times to be produced and to propagate. Responses involving preformed compounds are more rapid than those requiring synthesis of new chemicals or morphological responses. The induction of traits that provide defense against herbivores may not follow a simple or linear time trajectory. For example, soybeans that were attacked by Mexican bean beetles became more resistant after a lag of up to 3 days (Underwood 1998). Induction was relaxed 15 days after the initial attack, and after 20 days those plants that had been attacked became more susceptible to herbivores than uninduced plants. In other words, the stronger the induction of resistance was during the early days following attack, the stronger the induced susceptibility was later. Induced susceptibility presumably comes about as a by-product of other favorable traits.

There are relatively few systems for which the time course of induction and relaxation are well resolved. In general, relaxation lags following cessation of damage are much longer than induction lags. The cost of having inadequate defenses is presumably much larger than the costs of having unnecessary defenses. This discrepancy has been observed for systems that induce relatively rapidly and those that induce relatively slowly. For example, *Trifolium repens* mounted a systemic defense 38–51 hours (2–3 days) after caterpillar attack but required at least 28 days before it was relaxed (Gomez et al. 2010). Similarly, *Acacia drepanolobium* produced larger spines within 2 months of browsing by large mammals (Young and Okello 1998). However, relaxation for trees protected from browsing occurred gradually over more than 5 years (Young et al. 2003). Essentially all models that attempt to link induced responses to herbivore populations conclude that the timing of induction and relaxation are critical parameters that determine the consequences of the induced changes (Karban and Baldwin 1997:159–160).

6.2   Volatile signals and communication between ramets and individuals

Since all plants are sectorial to some extent, systemic communication among repeated organs by vascular connections is limited (section 6.1.4). Plant regions that are connected by vascular strands can exchange nutrients, photosynthates, hormones, etc. more readily than regions that do not share plumbing (fig. 6.1). In addition, risk of cavitation and drought stress limits the conditions under which vascular communication is profitable. One potential solution to limitations imposed on vascular signaling is the use of volatile signals to coordinate responses among nearby tissues within an individual plant (Farmer 2001, Karban et al. 2006, Heil and Adame-Alvarez 2010, Heil and Karban 2010).

6.2.1 Advantages and disadvantages of volatile communication

There are several potential advantages of a volatile communication system for signaling between tissues for an individual. (1) Volatiles are released immediately following damage by herbivores. (2) They are likely to move more rapidly than a vascular signal. (3) They can allow spatially close tissues to communicate even if those tissues do not share vascular connections to exchange information (Orians 2005). For some plants, such as lianas and trees, neighboring leaves on different stems may be attacked sequentially by feeding herbivores despite lacking vascular connections (Heil and Adame-Alvarez 2010). Phloem transports materials towards photosynthetic sinks and xylem transports materials towards actively transpiring tissues. Volatiles provide cues to leaves and other tissues that do not necessarily fall into one of these two categories.

There are also disadvantages to using a volatile system for communication. Once the volatile compound has been released, the emitter completely relinquishes control of the signal's fate. Volatile signals become public information while vascular signals remain private (Gershenzon 2007). Arrival at any receiver depends on wind turbulence, which means that the information may be lost, or it may be used by other competing plants, receptive herbivores, or other plant parasites. Despite these disadvantages, volatile communication has been found in many plants (see section 6.2.2).

Volatile signals can be perceived only over relatively short distances. Sagebrush individuals exposed to air from experimentally clipped conspecifics became more resistant to herbivores as long as those clipped emitters were within 60 cm; beyond this distance, no responses were detected (Karban et al. 2006). At a field site with wind that blows consistently in one direction, communication was detected only in branches that were downwind of the clipped branch (Karban et al. 2006). *Nicotiana attenuata* responded to sagebrush volatiles over even shorter distances, not exceeding 15 cm (Karban et al. 2000). Lima bean shoots responded to experimental induction over distances of 50 cm; over 80% of leaves belonged to the induced individual at this distance (Heil and Adame-Alvarez 2010). Beetle damage to alder trees increased as the distance between experimentally clipped trees and assay trees increased; this result was seen over distances up to 10 m (Dolch and Tscharntke 2000). These effects, measured in terms of plant damage, corresponded with changes in the number of herbivores that were attracted to trees at different distances from the experimentally clipped tree (Tscharntke et al. 2001). In summary, communication becomes less effective as the distance between emitter and receiver increases. As a consequence, limited transmission distances probably reduce the likelihood that unrelated individuals can eavesdrop on volatile signals.

If neighbors compete for resources, individuals may suffer if they provide information to unrelated competitors. Plants may reduce the costs of providing this information to eavesdropping neighbors by communicating selectively with kin. Experiments with sagebrush indicated that communication was more effective at reducing herbivore damage when emitter and receiver plants were genetically identical than when they were genetically different (Karban and Shiojiri 2009). Similarly, plants that were more closely related communicated more effectively than those that were unrelated strangers (Karban et al. 2013). The volatiles emitted by clipped plants were highly variable, although relatives were more likely to have more similar volatile profiles. It is not known whether (1) selection favors individuals to respond to their own volatiles and relatives have more similar volatile profiles and hence communicate more effectively, or (2) selection favors individuals that respond to cues from other plants that are more likely to share risk factors.

Another possible disadvantage to volatile communication following damage is that other herbivores may use those volatiles to locate suitable hosts. There is some evidence that plant volatiles can be used by herbivores to locate hosts and to stimulate feeding. Herbivores could reasonably use the volatiles that are released for plant communication although this has not been conclusively demonstrated. For example, herbivorous spider mites were attracted to volatiles emitted by cucumber plants infested with conspecifics but avoided plants that hosted thrips, omnivores that eat mite eggs (Pallini et al. 1997). Bean leaves that were lightly infested with spider mites emitted volatiles that were attractive to other mite individuals (Horiuchi et al. 2003). Heavily infested bean leaves emitted volatiles that repelled settling spider mites. It is not known how commonly herbivores use volatile cues of damage to locate hosts or as feeding stimulants. There are far more examples in which herbivore-induced volatiles act as cues that attract insects that are predators and parasites of the herbivores. However, if these carnivores have evolved receptor systems that respond to volatiles induced by herbivores, it is not unreasonable to imagine that herbivores can also respond to these cues.

Several volatiles have been shown to act as cues or signals that mediate eavesdropping or communication between plants (see section 4.3). For example, methyl salicylate and methyl jasmonate are volatile esters that may serve as long-distance signals; both are converted to more active acids (SA and JA) and other conjugates after being taken up by plant tissues (Shulaev et al. 1997, Park et al. 2007, Tamogami et al. 2008). Less is currently known about perception and transport of green leaf volatiles, although they have been implicated in communication in several plants (see section 4.3.8).

6.2.2   Empirical evidence for volatile communication

For some plants, communication has been demonstrated only between branches with direct vascular connections (see section 6.1.4). However, induced resistance for other species was as strong and as rapid in leaves that lacked vascular connections to the damaged leaf (Mutikainen et al. 1996, Kiefer and Slusarenko 2003). These later results suggest that the active signals are not limited to vascular channels. More direct support for volatile communication comes from studies that have blocked air movement. Sagebrush plants that had conspecific neighbors that were experimentally clipped became more resistant to herbivores (Karban et al. 2006). However, when air contact was reduced by clipping branches that were temporarily enclosed in plastic bags, no increases in resistance were observed in the neighbors. This was true both for within-plant communication among leaves on different branches of the same individual and for between-plant communication among leaves on branches of different individuals (Karban et al. 2006). Blocking root or soil contact had no effect on these results.

The first reports of volatile communication between plants that affected herbivores generated considerable controversy. In 1979, David Rhoades, a young scientist at the University of Washington, conducted several experiments to examine induced responses of red alders and Sitka willows (Rhoades 1983). He added caterpillars to the leaves of some trees, and had others protected from damaging insects, which he planned to bring to the lab to use as assays of caterpillar performance. However, the initial caterpillars that he added ate little and performed poorly, probably because they were diseased. To save his experiment, he reloaded caterpillars onto these trees and observed that they performed poorly on both previously damaged trees and previously undamaged controls. This result was not expected; he hypothesized that caterpillars on previously damaged trees would perform less well than those on undamaged controls. One possible explanation for these unexpected results was that damaged trees had released cues that made both the damaged individuals and their undamaged neighbors poor hosts for caterpillars. Rhoades conducted experiments over the next two seasons that produced results that were consistent with this hypothesis. He compared the performance of caterpillars on trees that were close to damaged neighbors to that of caterpillars on trees that were several kilometers away from damaged neighbors. All of the caterpillars grew poorly but caterpillars from the sites close to damaged trees lost more weight than those from the more distant control sites. There were no root connections between the damaged trees and their neighbors that had seemingly become more resistant to caterpillars. Rhoades interpreted these results as suggesting that volatile communication between damaged trees and their close neighbors was responsible for the induced resistance.

These results stimulated Ian Baldwin and Jack Schultz to conduct a series of lab experiments to examine volatile communication between potted poplar cuttings and maple seedlings (Baldwin and Schultz 1983). They placed plants that were damaged (torn) in one air-tight chamber and pumped air from these damaged plants into a second chamber containing assay plants. Assay plants that received air from damaged conspecifics were compared with another control group of assay plants that received air from a chamber containing no plants. Levels of phenolics were higher for plants with torn leaves than for controls. Plants that received air from conspecifics with torn leaves also had higher levels of phenolics than plants receiving air from the chamber without plants. They interpreted these results as indicating that volatile cues had been released from the torn leaves that were perceived by downwind assay plants, causing those downwind plants to increase production of phenolics. These results were published in *Science* and generated enormous attention from the popular press, which dubbed them "talking trees."

These initial reports were met with skepticism from the scientific community. First, other workers were unable to repeat Rhoades' observations of reduced caterpillar performance caused by volatile communication among alders (Myers and Williams 1984, Williams and Myers 1984). Rhoades himself was unable to produce consistent field results and was unable to get continuing grant support for this project (personal communication). Second, alternative mechanisms could have produced his results and these alternatives were not addressed by his study. Specifically, the caterpillars used in his experiments grew poorly and appeared to be diseased. Introducing caterpillars could have introduced infectious diseases to "damaged" trees and their close neighbors that were not introduced to trees at a more distant control site (Fowler and Lawton 1985). Third, the experimental designs used in both early studies had no true replication and therefore could not be used to evaluate whether some other factor or chance alone was responsible for causing the observed effects (Fowler and Lawton 1985). Rhoades had all of his replicates near to damaged trees at a single site and all of his replicates far from damaged trees at a second site. Baldwin and Schultz (1983) had all of their replicates of each treatment in a single chamber. In both of these experiments it was impossible to distinguish effects that were caused by the treatments from those caused by disease or another factor with statistical confidence (Hurlbert 1984).

An influential review paper convinced most biologists that the evidence supporting volatile communication that affected herbivores was weak or nonexistent (Fowler and Lawton 1985). The subject was dismissed as something that had been suggested and subsequently debunked, somewhat akin to cold fusion. The phenomenon regained credibility with a publication in 1990 by Ted Farmer and Bud Ryan, the latter a well-established biochemist and member of the National Academy of Sciences. They reported that volatile methyl jasmonate induced the production of proteinase inhibitor, a putative plant defense, in tomato leaves. Furthermore, incubating branches of sagebrush, which emit methyl jasmonate, with potted tomato plants caused the tomato foliage to accumulate proteinase inhibitor (Farmer and Ryan 1990). Several years later, a similar role was found for methyl salicylate as a signal for plant resistance to pathogens (Shulaev et al. 1997). During this decade several workers conducted well-replicated experiments that indicated that plants with air contact with clipped neighbors experienced less herbivore damage than those with air contact with unclipped neighbors, although these results were rejected repeatedly by journals until 2000 (Dolch and Tscharntke 2000, Karban et al. 2000).

In recent years evidence has accumulated that volatile communication can affect plant resistance to herbivores (Karban et al. 2014). A statistical meta-analysis of 48 well-replicated studies found that, overall, cues from damaged plants made neighbors more resistant. This result was not universal and there were cases in which cues from damaged plants made neighbors more attractive or susceptible to herbivores. This meta-analysis included studies in which cues from damaged neighbors affected plant traits (e.g., Farmer and Ryan 1990, Tscharntke et al. 2001, Engelberth et al. 2004), reduced insect performance (e.g., Tscharntke et al. 2001, Ton et al. 2007, Peng et al. 2011), and diminished levels of damage experienced by neighbors (e.g., Karban et al. 2000, 2006, Heil and Silva Bueno 2007). There was no indication that some response variables (plant vs. insect measurements) were more likely to be affected by volatile cues.

Several life history traits and environmental conditions were expected to make volatile communication more likely to be selected for or detected by experimental studies. Volatile communication was more likely to be found in greenhouse and lab studies of agricultural crops than in field studies of wild plants (Karban et al. 2014). Studies involving agricultural crops in laboratory environments presumably reduce background variation due to genetic heterogeneity and environmental noise and make subtle signals easier to detect.

It was expected that plants from arid environments might rely more heavily on volatile communication because they tend to be more sectored than plants with access to a persistent supply of water (Waisel et al. 1972, Zanne et al. 2006, Schenk et al. 2008). This expectation was not found in the meta-analysis (Karban et al. 2014). Similarly, woody plants tend to be more sectored and at greater risk of a rupture in the xylem stream (cavitation), which might favor volatile rather than vascular coordination. However, there was no evidence that woody species relied more on volatile communication than herbaceous species.

Many of the studies that reported volatile communication did not examine why reductions in plant damage occurred. In other words, it was rarely determined whether plants became less attractive to herbivores, less favorable for herbivore growth, etc. For several of these systems, there was evidence that volatile cues primed plants to respond more rapidly or effectively once they have been actually attacked (e.g., maize [Engelberth et al. 2004], wild tobacco [Kessler et al. 2006], wild lima beans [Heil and Silva Bueno 2007], poplar [Frost et al. 2008], *Vaccinium* [Rodriguez-Saona et al. 2009], cabbage [Peng et al. 2011]). It is possible that exposure to low concentrations of volatile cues primes plants while exposure to higher concentrations of cues induces resistance, although this hypothesis has not been rigorously tested.

Defenses that are induced only when needed are generally believed to be less expensive than constitutive defenses that are always employed (Karban and Baldwin 1997, Agrawal 2005). Priming may allow plants to be even more plastic and optimize their investments in defense since priming itself should be relatively inexpensive and induction can be reserved for situations when it will be beneficial. Although there have been relatively few studies that have examined the fitness consequences of responding to volatile cues, existing evidence suggests that plants benefit by eavesdropping on the volatile cues of neighbors and the volatile signals released when their own tissues have been attacked (Karban et al. 2012).

It makes intuitive sense that plants will be sensitive to environmental information that they can acquire and *respond* when that information is reliable and predictive. It is harder to understand how and if selection favors plants that *emit* volatile cues when they have been attacked.

6.3 Indirect defenses against herbivores

Plants defend themselves against herbivores not only by producing morphological and chemical hindrances, but also by communicating with and manipulating other organisms that provide protection from herbivores. The most common organisms that act as plant "bodyguards" are predators and parasites of herbivores (Price et al. 1980). Plants attract these bodyguards by providing either resources (food and shelter) or information about prey that allows these carnivores to forage more effectively.

6.3.1 Plants provide food and shelter to predators and gain protection from herbivores

Naturalists have been aware that plants provide shelter for predators for some time. The Swedish naturalist Axel Lundstrom proposed that many leaves have specialized structures—cavities, and hairs called domatia—that house predaceous and fungivorous mites and small insects (fig. 6.3a) (Lundstrom 1887). Surveys in deciduous temperate forests in eastern North America and Korea have revealed that these structures are present on approximately half of the woody species in these habitats (Willson 1991, O'Dowd and Pemberton 1994). Various experiments removing domatia or providing plants with additional domatia supported the hypothesis that these structures increase numbers of predators, decrease plant losses, and increase plant fitness (Walter 1996, Agrawal and Karban 1997, Norton et al. 2000, Romero and Benson 2005).

Domatia that house mites and small predaceous insects are more common for plants growing in temperate habitats, while plants that provide

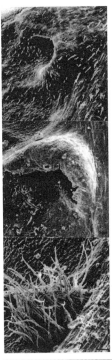

FIGURE 6.3 Domatia and associated rewards provided by plants to attract ants and other predators of herbivores and leaf fungi. A. Microscopic domatia house predaceous and fungivorous mites. Three kinds of domatia are shown: a leaf pouch, a leaf pocket, and a tuft of leaf hairs. From O'Dowd and Willson 1989. B. *Acacia drepanolobium* provides nectar to ants from extrafloral nectaries. Photo courtesy of Todd Palmer. C. *Acacia drepanolobium* also provides hollow swollen thorns that house the ants. Photo courtesy of Todd Palmer.

shelter and food resources for ants are more common in the tropics. Approximately one third of all woody plant species provided food in the form of extrafloral nectar for ants in a Panamanian forest (Schupp and Feener 1991) (see fig. 6.3b). Experimental removal of ants demonstrated their importance as bodyguards that protect their associated plants from herbivores, encroaching vegetation, and fungi (Janzen 1966, 1967, Rico-Gray and Oliveira 2007).

Plants that rely on ant defenders provide cues and additional resources to these defenders when the plants are attacked. For example, *Cecropia* trees that were experimentally damaged emitted volatile cues that rapidly recruited *Azteca* ants that defended them (Agrawal 1998b, Agrawal and Dubin-Thaler 1999). Local recruitment of ants was proportional to levels of local damage, and leaves that had been damaged previously recruited more ants than those damaged for the first time. Ants also responded to cues emitted by damaged neighbors to increase recruitment. *Cecropia* leaves near experimentally damaged neighbors accumulated more ants than leaves near undamaged neighbors (Agrawal 1998b).

Plants provide food for ants and other carnivores as sugar in extrafloral nectar and as protein and fat in "food-bodies," which have been less studied. Extrafloral nectaries were described by various 18th- and 19th-century naturalists; extrafloral nectaries are often accompanied by structures that specifically house ants (Bentley 1977). Extrafloral nectar is produced by leaves, shoots, and inflorescences of many plants but is not involved in attracting pollinators (Koptur 1992). Plants with extrafloral nectaries increased nectar volume or improved nectar composition following attack; these changes were associated with greater ant recruitment (Agrawal and Rutter 1998). The production of extrafloral nectar was correlated with the intensity of attack by herbivores (Heil et al. 2001). Furthermore, the intensity of damage was positively related to endogenous levels of JA, and experimentally supplied JA increased nectar production. Extrafloral nectar in damaged leaves had increased concentrations of sugars (Stephenson 1982) and amino acids (Smith et al. 1990), two potentially important resources for ants (and herbivores).

It is interesting to note that the same compound, JA, is used as a signal to induce both direct physiological responses to herbivores and to supply resources in order to increase indirect defenses involving ants. Volatiles from damaged neighbors primed extrafloral nectar secretion of bean plants (Choh and Takabayashi 2006, Heil and Silva Bueno 2007). In this case, bean leaves exposed to volatiles from damaged neighbors produced more extrafloral nectar when they were later attacked than leaves that not been previously primed.

Plants that were exposed to cues indicating a high risk of herbivory increased their investment in structures and food rewards that attract predators of their herbivores. Conversely, plants that experienced reduced risk of herbivory reduced their investments in both shelters and food for ants. *Acacia* trees that were protected from mammalian herbivores for one year produced as many swollen thorns that house ants (fig. 6.3b,c) and extrafloral nectaries as plants that experienced herbivory, but those protected for 7 years produced fewer shelters for ants and fewer nectaries than plants that experienced ambient levels of browsing (Huntzinger et al. 2004). Relaxation of these rewards following removal of mammals made trees more vulnerable to beetle attacks, and reduced tree growth and survival (Palmer et al. 2008).

Many studies indicate a positive relationship between plant resources, attraction of ants, and plant defense. Ants are effective defenders of plants because they are effective foragers, recruit to sources of food and other resources, and defend these resources from competitors and other threats such as herbivores or plants that might compete with their resource provider. Many studies have excluded ants from plants with extrafloral nectaries and found that plants without their ants experience higher rates of herbivory (reviewed by Bentley 1977, Koptur 1992, Heil 2008). Interpretation of these results is complicated because the experimental treatments also reduced numbers of crawling herbivores. Application of JA increased secretion of extrafloral nectar, increased ant visitation, and decreased leaf damage, although these treatments may have induced other defensive responses as well (Heil et al. 2001).

Many other visitors, including herbivores, exploit extrafloral nectar as well as other plant foods and shelter, and these species are not necessarily mutualists with plants. For example, *Acacia drepanolobium* at one study site is commonly inhabited by any of four ant species, one of which resides in its thorns and consumes its extrafloral nectar, but provides little defense, and another that sterilizes its flowers (Stanton et al. 1999, Palmer et al. 2008). However, this suite of ant species appears to provide a net fitness benefit to plants when considered as an entire community (Palmer et al. 2010). In most of these interactions, the plant that benefits by being defended and the ants that benefit by exploiting resources that the plant provides have conflicting interests.

6.3.2 Plants provide information to predators about herbivores

Plants also communicate with predators and parasitoids of their herbivores when they are attacked. Unlike the systems described above, in which the plants provide valuable resources to these carnivores, here the plants provide reliable information that presumably enables the predators and parasites to find prey more efficiently. While food and shelter that are provided constitutively are useful to predators, information is valuable only if it is current and reliable (Kessler and Heil 2011). The idea that plants provide information that attracts predators and parasitoids of its herbivores is relatively recent. Lima bean leaves that were attacked by spider mites released volatile signals that attracted predatory mites (Dicke and Sabelis 1988). Plants remained attractive to predators for several hours after herbivory (Sabelis and van de Baan 1983). Maize seedlings that were being fed upon by caterpillars emitted volatile terpenoids that attracted parasitic wasps (Turlings et al. 1990). Female wasps used the terpenoids to locate and parasitize feeding caterpillars. Parasitoids are noted for their ability to learn various reliable cues that they then associate with the presence of suitable hosts. More recently, many other workers have observed that predators and parasites used plant-derived volatiles to locate prey; these included entomophagous nematodes that responded to insect-induced sesquiterpene cues that traveled through the soil (Rasmann et al. 2005), and birds that increased attack rates on trees that were emitting cues caused by caterpillar feeding (Mantyla et al. 2008).

Herbivory causes essentially all plants to modify their volatile profiles, including the identity and relative emission rates. Many of these emissions are likely the result of inevitable and general processes, although they still may provide reliable information for carnivores. For example, wounding causes the breakdown of chloroplast membranes, releasing linoleic acid, and associated green leaf volatiles, common to all plants (Noordermeer et al. 2001). In response to herbivory, many plants produce an additional blend of volatiles dominated by compounds that are not found in the emissions of undamaged or mechanically damaged plants (Dicke and van Poecke 2002). Herbivore-induced volatiles are emitted systemically, from both damaged and undamaged plant tissues (Turlings and Tumlinson 1992, Dicke et al. 1993). Plant-derived volatiles are more attractive to predators and parasitoids than volatiles from the herbivores or their frass (Sabelis et al. 1984, Turlings et al. 1991, Turlings and Wackers 2004). Plants also appear to become primed by previous damage and to increase emissions of attractive volatiles following subsequent attacks (Ton et al. 2007).

The majority of the volatile compounds emitted by damaged plants are either fatty acids produced through the lipoxygenase pathway, terpenoids produced through the isoprenoid pathway, or phenylpropanoids produced through the shikimic acid pathway (Dicke and van Poecke 2002, Dudareva et al. 2006). Damaged plants commonly release multiple volatiles (20–200 compounds) in amounts that can be detected by existing technology, and establishing which components of the blend are biologically effective has been difficult (Dicke and van Loon 2000, Dudareva et al. 2006).

Volatile emissions show some specificity that depends upon the plant species and genotype, at least in terms of the relative frequencies of the various compounds (Dicke and van Poecke 2002, Hare 2011). Volatile blends may also depend upon the particular herbivore causing the damage, involving different ratios of the same compounds (Dicke 1999). Different instars of the same herbivore species cause different emissions in some instances (Takabayashi et al. 1995). Parasitoids recognized the volatiles emitted in response to their host, *Helicoverpa virescens*, in preference to those emitted in response to a nonhost, *Helicoverpa zea* (De Moraes et al. 1998). In several cases, plant emissions were specific to particular elicitors in the salivary secretions of their herbivores (Mattiacci et al. 1995, Alborn et al. 1997). Specificity of cues provides predators and parasitoids with specific information about the herbivores causing the damage. However, not all responses of carnivores to plant volatiles showed fine-tuned specificity, and some predators and parasitoids were attracted to cues of hosts that they were unable to exploit (Agrawal and Colfer 2000, Thaler et al. 2002).

Some parasitoids and predators have the ability to learn to associate particular plant volatiles with prey (Allison and Hare 2009). This is particularly well documented for parasitoids that respond to different herbivore-induced volatiles (Lewis and Tumlinson 1988, Roitberg et al. 1993, Turlings et al. 1993, Geervliet et al. 1998a, Geervliet et al. 1998b). Predatory mites also have the ability to associate plant volatiles with the presence of prey (de Boer et al. 2005, Takabayashi et al. 2006). Under certain conditions, learning enables parasitoids and predators to hunt more successfully and increase their fitness (Dukas and Duan 2000).

Responses of carnivores to herbivore-induced plant volatiles are common and relatively well studied. In contrast, we know far less about responses of herbivores to these volatiles. In general, there are more reports of herbivores being attracted to emissions caused by feeding of conspecifics than reports of repellence (Dicke and van Loon 2000). Most of these studies were conducted in laboratory settings, and herbivores responded to plant quality rather than mating opportunities. Herbivore responses can be conditional; for example, spider mites were attracted to plants hosting low densities of conspecific mites but repelled by plants with high densities of potential competitors (Dicke 1986, Horiuchi et al. 2003). They were strongly repelled by plants with thrips, omnivores that feed on both vegetative tissue and mite eggs (Pallini et al. 1997). Since information emitted by plants is released into the environment, diverse species may eavesdrop and respond to these public cues. For example, freshwater macrophytes emit cues that attract herbivorous snails; the snails feed on epiphytes that cover the macrophytes and compete with them for light and nutrients (Bronmark 1989).

Carnivores are known from the field to use herbivore-induced plant volatiles to locate prey (e.g., Drukker et al. 1995, Thaler 1999a), but this area is particularly lab-based, and field verification is often currently lacking. Pear trees with neighbors that were harboring psyllids in cages attracted more predators than trees without neighbors with herbivores (Drukker et al. 1995). Predators responded to changes in psyllid densities; blocking the volatile cue with plastic sheets resulted in a drop in predator numbers. Caterpillars near tomato plants treated with JA experienced higher rates of parasitism than caterpillars near control plants (Thaler 1999a). Wild tobacco plants treated with MeJA and those with simulated volatile emissions attracted more egg predators and repelled herbivores, reducing caterpillar numbers by 90% (Kessler and Baldwin 2001).

Predators and parasites may reduce prey but the evidence that emissions of herbivore-induced volatiles benefit plants is equivocal (Allison and Hare 2009, Kessler and Heil 2011). A convincing demonstration requires that herbivores reduce plant fitness, that predators or parasites reduce this negative effect, and that the carnivores be attracted by herbivore-induced volatiles. Each of these requirements has been satisfied, although no single system provides convincing evidence of all three. The case is stronger for predators than for parasitoids. Predators kill their herbivore prey, preventing the prey from any further feeding on the host plant. In contrast, parasitized herbivores continue to feed on the host and may actually increase their consumption relative to unparasitized individuals (Slansky 1978, Coleman et al. 1999). Surprisingly, a recent meta-analysis indicated that parasitoids were more likely to provide a large benefit to the host plant than were predators (Romero and Koricheva 2011). Predators may also have other negative consequences that diminish their beneficial effects to plants as consumers of herbivores, e.g., repelling or killing pollinators (Knight et al. 2006, Willmer et al. 2009).

In general, plants that provide carnivores with information benefit less from the interaction than plants that provide rewards in the form of food and shelter (Romero and Koricheva 2011). The evidence that plants increase their fitness by providing reliable information to carnivores is suggestive but not conclusive. *Arabidopsis* plants that were fed upon by parasitized caterpillars produced more seeds during their lifetimes than those fed on by unparasitized caterpillars (van Loon et al. 2000). Similarly, maize plants that were attacked by parasitized caterpillars produced approximately 30% more seeds than plants attacked by unparasitized caterpillars (Fritzsche Hoballah and Turlings 2001). Both of these studies started with experimentally parasitized caterpillars in very controlled circumstances; neither demonstrated that parasitoid attacks were mediated by herbivore-induced volatiles.

Predators may be more likely than parasitoids to kill herbivores and therefore to reduce damage to plants. In some agricultural situations, spider mites can devastate crop plants so that attracting predators is presumed to allow plants to survive and reproduce (Dicke and Sabelis 1988). Spider mites rarely reach high densities in nature, so extrapolating from these situations to more natural ones is risky. Wild tobacco plants attacked by their herbivores emitted volatiles that attracted generalist predatory bugs (Kessler and Baldwin 2001). These bugs reduced the survival of feeding caterpillars, although plant fitness was not measured in these experiments. In summary, it is possible that herbivore-induced volatiles may benefit plants by attracting the predators and parasites of their herbivores, although this has not been demonstrated conclusively in any system and alternative explanations have not been carefully evaluated (Hare 2011).

6.4   Visual communication between plants and herbivores

Most of this chapter has focused on biochemical cues used for communication between plants, their herbivores, and the natural enemies of those herbivores. This makes some sense since the majority of herbivores and their natural enemies are insects, and insects rely on chemical cues to locate food and to communicate (Bradbury and Vehrencamp 1998). However, workers studying plant defenses have historically been fascinated with chemical adaptations (e.g., Fraenkel 1959, Ehrlich and Raven 1964), and other important forms of defenses and interactions are relatively understudied (Carmona et al. 2011). Many insect and vertebrate predators also use visual cues to locate and discriminate among potential host plants.

Plants match their visual (and chemical) backgrounds to reduce the probability that they will be detected by herbivores. A well-known example involves *Lithops* spp. that look very much like the stones with which they share the desert floor (fig. 6.4a) (Wickler 1968, Wiens 1978). The hypothesis that this plant gains protection from crypsis is plausible but has not been tested. One system that has been investigated experimentally is the non-photosynthetic parasitic plant *Monotropsis odorata*, which is difficult for a human eye to distinguish from the leaf litter that it grows among (Klooster et al. 2009). This crypsis is achieved by a covering of dried, brown bracts; removing these bracts caused plants to experience 20–27% more herbivory and 7–20% less fruit production. Leaves of the vine *Boquila trifoliolata* match their woody support in terms of size, color, and morphology, and vines that used multiple hosts produced the appropriate leaves to mimic each host (Gianoli and Carrasco-Urra 2014). Leaves of vines suffered less herbivory when they matched their host than when they didn't. The mechanisms that allowed the vines to sense and respond to their supports are not yet known.

Seeds of many plant species closely match the color of the background soil in their local environment. For example, the large-seeded legume *Acmispon wrangelianus* grows on a variety of local soil types ranging in color from brown to gray to green (Porter 2013). The color of the seeds closely matches the color of the background soil in any given place despite the opportunity for gene flow among populations on different soil types. In other systems, such as *Pinus sylvestris* and *P. halepensis*, seeds that matched the background color of their environments were less likely to be consumed by birds (Nystrand and Granstrom 1997, Saracino et al. 2004).

Plants may be aposematic, advertising to herbivores that they bear toxins, thorns, or spines or are otherwise unpalatable (Lev-Yadun 2009). Many thorns, spines, and prickles are conspicuously white or colorful or are associated with noticeable spots or stripes. Presumably herbivores learn to avoid these conspicuous and sharp tissues since even herbivores like peccaries, which are adapted to feeding on spiny plants, prefer tissues that are less spiny (Theimer and Bateman 1992). I am not aware of any experimental tests that have evaluated whether the hypothesized warning colors actually deterred herbivores. Seeds have also been hypothesized to be aposematically colored (Lev-Yadun 2009). Seeds of *Eremocarpus setigerus* are mottled and cryptic when they contain low levels of toxins and conspicuous plain gray when they are more toxic. Mountain doves rejected the gray seeds in favor of mottled ones (Cook et al. 1971).

Plants may also mimic insect eggs and damage in an attempt to discourage conspecific insects from ovipositing or otherwise settling. Plants with

FIGURE 6.4 Plants mimic other objects to gain protection. A. *Lithops terricolor* resembles rocks in its environment. Photo courtesy of Rob Skillin. B. Leaves of *Caladium steudneriifolium* have variegated patterns (right) that resemble insect mines (left). From Soltau et al. 2009.

structures that closely resembled insect eggs or larvae may receive fewer actual eggs (Benson et al. 1975, Shapiro 1981). Removal of the egg mimics produced by the plant resulted in greater rates of oviposition (Shapiro 1981). Other workers have hypothesized that plants have structures that mimic ants, aphids, and caterpillars and that these patterns deter actual attacks by herbivores (Lev-Yadun 2009). For example, the leaves of *Caladium steudneriifolium* sometimes have white markings that resemble the mines made by a caterpillar that commonly attacks them (fig. 6.4b) (Soltau et al. 2009). Leaves without these markings had 4–12 times as many actual mines; green leaves that were experimentally painted to resemble the markings received ½₀ the number of attacks. The bright colors of many fungi have been interpreted as aposematic, and plants hosting fungi that are highly toxic to vertebrates may benefit through association with these fungi that herbivores avoid (Lev-Yadun and Halpern 2007). These reports are plausible and stimulating although hypotheses that plants escape herbivory by being cryptically or aposematically colored or by mimicking herbivore signs have not yet been rigorously evaluated.

Plant-herbivore interactions have been studied intensively over the past few decades and we have learned a lot about how plants perceive herbivores and respond to defend themselves. Plants respond with changes in membrane potential and ion fluxes, and levels of hormones. Damage by herbivores can offer reliable information about risk of future attack. Volatile cues emitted by damaged plants provide information to neighboring tissues on the same individuals, other individual plants, herbivores, and the predators and parasitoids of herbivores.

7 Plant Communication and Reproduction

Many plants are unable to reproduce without the help of external agents. Outcrossing will not occur without wind, water, or animals transferring pollen from the stamens of one individual to the stigma of another receptive individual of the same species. Seeds will not germinate in favorable new sites without the aid of wind, water, or animals moving them from their parent to this new location. Animals that transfer pollen or seeds are often considered to be involved in mutually beneficial interactions with plants by providing reproductive services. In this chapter we will look at the rewards that plants provide, most commonly in the form of nectar, pollen, and nutritious fruits and seeds. We will also examine the cues they provide to attract pollinators and seed dispersers including morphology, color, visual patterns, and scents. Female plants also use information to influence reproductive outcomes, screening pollen donors and differentially allocating resources to more favored offspring.

It is important to keep in mind that the goals of plants and their animal vectors are actually quite different. The plant has been under selection to move its pollen and seeds to those locations where they will be successful in the sense of producing the next generation of reproductive plants. The plant has been under no specific selection to nourish the animals that provide these services. In some cases, plants essentially sacrifice many of their potential offspring (seeds) so that some small fraction of offspring can survive (think acorns and squirrels); the chance of offspring survival may be enhanced if something unfortunate happens to its animal partner. The squirrel that dies or forgets about a cache of acorns is probably more helpful to the parent plant than the one that succeeds in nourishing itself. Conversely, the animal vectors have been under selection to collect and digest resources that plants produce. Pollination and seed dispersal are secondary consequences from their point of view. To accomplish their different objectives, plants and animals communicate using cues and signals that provide both honest and deceptive information.

7.1 Pollination and communication

Animals most commonly visit flowers to collect pollen and nectar that they then consume or collect for their offspring to consume. Most animal visitors are insects, but various invertebrates, birds, bats, lizards, and others also seek out these resources. Pollen sticks to the bodies of these visitors and may be deposited on receptive stigmas of other flowers that the visitor subsequently contacts. Any animal that visits flowers is likely to be a pollinator to some extent. Even crab spiders that hunt for pollinating insects at flowers move some pollen between flowers. Some visitors are far more effective as pollinators than others, and plants may attempt to limit the use of their resources to these effective vectors.

7.1.1 Flowers offer rewards to visitors

Plants provide pollen, nectar, and other rewards to the animals that visit their flowers and, by visiting, pollinate them. Plants generally present dense concentrations of pollen grains on anthers, which facilitates movement by pollinators. However, pollen is rich in protein, and the animals' goal is to harvest the pollen for consumption. That is, pollen serves two distinct functions—plant reproduction and visitor reward—and these two are mutually incompatible (Willmer 2011). Flowers that use pollen to attract visitors must produce large quantities since visitors often consume most of it. Some plants that successfully reward visitors with pollen are able to place some of the pollen on the visitor's body in a place where it cannot be consumed. Other plants that rely on visitors that harvest pollen produce two types of pollen—brightly colored pollen that is mostly consumed and inconspicuous pollen that is more likely to fertilize the ovules of conspecifics

(Faegri and van der Pijl 1979). These two types of pollen are often spatially segregated on different anthers or flowers.

Nectar has become the main reward for flower visitors over evolutionary time, allowing plants to save their pollen for reproduction (Willmer 2011). Nectar is rich in carbohydrates and is secreted by specialized glands at the base of flowers, and sometimes from other organs (extrafloral nectar, as discussed in section 6.3.1). Most nectars are rich in sucrose, glucose, and fructose, although plant species vary in the concentrations of sugars, amino acids, lipids, and secondary chemicals in their nectar (Baker and Baker 1975). Although there are some exceptions (see discussion of toxic nectar in section 7.1.3) (Adler 2000), nectar feeders tend to be generalists; nectar is consumed by any animal that can access it. Nectar is generally replenished during the night and depleted by nectar feeders during the day (Faegri and van der Pijl 1979). Current stores of nectar rewards are often not visible to animal visitors before they sample each flower, making expected nectar rewards uncertain (Schaefer and Ruxton 2011). Pollen rewards are more visible and more predictable (Faegri and van der Pijl 1979). Visitors are often attracted directly to visible pollen but rely on indirect cues of inconspicuous nectar rewards.

Natural selection can favor flowers that offer unusual rewards as long as those rewards attract visitors that move pollen. Flowers of some species are thermogenic and may offer insects, which are ectothermic, a shelter that is warmer than the surrounding environment (Meeuse 1975, Schneider and Buchanan 1980, Seymour and Schultze-Motel 1996, 1997). Flowers of other species move to track sunlight or retain warmer temperatures by other means that make them valuable habitats for insect visitors (Kevan 1975, Sapir et al. 2006). Some flowers provide resins and volatile chemicals that visitors can collect and consume, attract mates with, and use to build nests (Dodson et al. 1969, Dressler 1982, Ramirez et al. 2011).

7.1.2 Flowers advertise for pollinators

Female fitness in natural plant populations is commonly limited by failure to receive sufficient pollen (Knight et al. 2005). As a result, plants compete for the services of pollinators and communicate with potential visitors by what we could call advertising (Robertson 1895, Mitchell et al. 2009). The most widespread advertisements involve flower morphology, color, and other visual patterns, as well as floral scents.

As a rule, larger flowers and floral displays are more attractive to pollinators than smaller ones (Galen 1999). Some visitors are attracted to radiating and symmetrical shapes and patterns (Lehrer et al. 1995). The colors of flowers make them more or less attractive to particular kinds of visitors—red is more attractive to birds, blue and yellow to bees, white to moths, and so on (Fenster et al. 2004, Willmer 2011, Rosas-Guerrero et al. 2014). These "pollination syndromes" have many exceptions and the utility of these generalizations has been challenged, since most plants use a variety of pollinating agents (Waser et al. 1996, Ollerton et al. 2009). Plants that have their pollen transferred by many different visitors are less dependent on the changing spatial and temporal distributions of individual visitor species.

Many visitors are attracted to flowers that present strong visual contrasts with their surroundings (Faegri and van der Pijl 1979). The preference of bees for contrasting colors is so strong that crab spiders can use it to their advantage (Heiling et al. 2003). The spider *Thomisus spectabilis* contrasts sharply with the flowers within which it sits and waits. Bees are attracted to the contrasting pattern, and they visit flowers with spiders in preference to those without them, with deadly consequences.

In addition to bright or contrasting base colors, many flowers attract visitors with complex visual patterns. Nectar guides were originally described by Christian Konrad Sprengel in 1793 (see Sprengel 1996 for an English translation) and the name has persisted, although "nectar guides" are also associated with pollen rewards (Lunau 2000). Floral guides are often concentric rings around the reward, or radiating lines pointing towards the reward; they generally contrast with the background color and have high color purity (see fig. 2.3). In experiments with artificial flowers, the presence of nectar guides enabled bees to discover the reward more quickly and reliably, which should benefit both the bees and the plant (Leonard and Papaj 2011). However, when the flowers no longer possessed a reward, bees were still more likely to visit these flowers with now dishonest nectar guides. South African irises that had their flower guides experimentally painted over experienced reduced probing by fly pollinator visitors, pollen removal, and fruit set (Hansen et al. 2012).

Flowers also produce volatile scents that attract visitors. Unlike pollen, scents are rarely rewards in themselves but visitors can associate them with rewards. Floral odors may be extremely complex blends exceeding 100 compounds that act in a highly context-dependent manner and are capable of causing diverse consequences (Raguso 2008). For example, floral emission of MeSa attracts orchid bees but also predatory arthropods, attenuates visits by honeybees, and affects mating of *Pieris* butterflies. The same compound can even have different effects on the same visitor species depending on the context. For example, male cycad cones emit volatiles in the relatively cool mornings that attract pollen-feeding thrips (Terry et al. 2004, Terry et al. 2007). During midday, male cones heat up, causing a 1-million-fold increase in the concentration of these compounds, repelling the thrips. At this time, thrips are attracted instead by female cones, which emit these same compounds at only $\frac{1}{5}$ the concentration. Thus the context-dependent cues increase the likelihood of successful pollen transfer from male to female cycads. Visitors such as honeybees perceive subtle differences among flowers, and their choices involve both concentration intensity and ratio of volatile compounds (Wright et al. 2005). Often insect visitors can learn to associate rewards more effectively with scent than with other cues, and scent learning enhances the likelihood that individual visitors will specialize (Menzel and Muller 1996, Wright and Schiestl 2009).

7.1.3 Not all flower visitors are pollinators

Plants benefit only when visitors move pollen from the anthers of one individual to receptive stigmas of another individual of the same species. A visitor that deposits pollen on the stigma of another species is no more useful than one that consumes the pollen. Visitors can damage the reproductive organs and introduce venereal diseases (Kaltz and Shykoff 2001, Morris et al. 2010). A stigma that receives heterospecific pollen is more likely to become blocked or clogged and less receptive to pollen of the appropriate species (Thomson et al. 1982, Morales and Traveset 2008). Therefore, plants attempt to advertise for visitors that will be motivated to visit other flowers of the same species. Providing a reward that is too large will not motivate the visitor to seek another conspecific individual, nor will a reward that is too small.

Animals that visit flowers of the same species will be more valuable as pollinators than those that visit many species and alternate among them (Darwin 1876). Visitors are said to be flower constant, meaning that individuals specialize on one or a few plant species (Plateau 1901, Waser 1986). Flower-constant individuals will pass over flowers of other species even when these flowers offer rewards of equal or greater value (Heinrich 1979). Different individuals of the same visitor species may visit different plant species. Flower constancy may be enhanced by the formation of transient search images, but be limited by short-term memory of cues or the motor memory required to retrieve rewards from complex flowers (Darwin 1876,

Chittka et al. 1999, Goulson 2000). In addition, flower constancy improves efficiency by reducing time spent sampling alternatives.

Plants may increase their likelihood of being visited by flower-constant pollinators by restricting access to their flowers by less effective pollinators. Plants can limit visitors with morphological restrictions such as fused corolla tubes that are accessible only to specialized long-tongued insects and birds.

Plants also restrict visitation by producing cues that appeal to the particular sensory biases of some limited range of visitors. Some flower visitors possess innate preferences for particular rewards and for particular floral cues, such as colors or scents. For example, naïve, newly emerged bees that have never encountered flowers have a strong bias for certain colors (Giurfa et al. 1995, Lunau et al. 1996). Visitors also learn to orient toward cues that indicate the presence of rewards (Raguso 2008). For example, adult moths learned to prefer individual floral volatiles when they were associated with increased rewards (Cunningham et al. 2004, Cunningham et al. 2006). This is significant since flowers of different species often emit similar compounds but in different concentrations. These differences produce unique volatile signatures and provide cues to visitors that are far more species-specific (Dudareva et al. 2006). Studies indicate that the presence of multiple cues (e.g., color along with odor) that differentiate flowers of different species increases flower constancy relative to single cues (Gegear and Laverty 2005, Leonard et al. 2011).

Plants are able to take advantage of the instincts of animals, to feed and to reproduce. In some instances, plants deceive their visitors, attracting them with dishonest cues but providing no reward (Renner 2006). Visitors may be unaware of the lack of rewards or may act instinctively following a chain of behaviors that are adaptive under other circumstances. For example, *Ophrys insectifera* flowers look strikingly like female wasps and attract males of that species (Wolff 1950). The flower mimics the size, shape, and texture of the female including "eyes" and "antennae." As a male attempts to copulate with the flower, his head and abdomen contact pollinia, which he carries off. Males attempt to mate repeatedly and in so doing deposit the pollinia on the stigma of another *O. insectifera* individual. Many other orchid species employ similar mimetic tricks to entice males to "copulate" with them (fig. 7.1). In addition, other plants produce flowers that some male insects chase or sting when they mistake them for conspecifics or prey. The flowers are pollinated as a result of the aggressive behaviors of these insects, which receive no reward in return (Faegri and van der Pijl 1979).

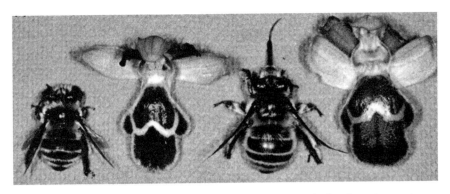

FIGURE 7.1 Male anthophorid bees pollinate orchids that they mistake for females and attempt to copulate with. The photograph shows the female bees and their orchid mimics. From left to right: *Anthophora sicheli* (female bee) and *Ophrys fleischmanii* (orchid); *Anthophora dalmatica* (female bee) and *Ophrys omegaifera* (orchid). From Schluter and Schiestl 2008.

Once flowers have been fertilized they may quickly become less attractive—wilting, closing, or abscising petals and stigmas, and reducing the production or apparency of other attractive traits such as odor, nectar, or nectar guides (Gori 1983, Weiss 1995, van Doorn 1997). Producing expensive attractants after pollination is a liability, expending resources, increasing the likelihood that visitors will damage reproductive tissues, and possibly causing other unfertilized flowers to be less likely to be visited. For example, bright yellow *Oenothera drummondii* were attractive to visitors but orange-red flowers were not (Eisikowitch and Lazar 1987); experimental treatments involving touching the reproductive organs or withdrawing nectar caused flowers to close and become less visually attractive. For *Lupinus pilosus*, pollen release (male function) stimulated small changes in petal color although pollen deposition on the stigma or growth of pollen tubes (female function) acted as more effective triggers for color change (Nuttman and Willmer 2003). Changes in attractiveness after pollination can act as honest signals directing visitors towards flowers with rewards and towards flowers that have not yet been pollinated.

Plants sometimes take a more active role in making visitors into pollen transporters. Some orchid flowers can sense that a visitor is near and shower it with pollen (Jaffe et al. 2002). Several species can sense the vibrations from the wings of nearby bees. The flowers then fling sticky pollinia onto the bee, or otherwise cover it with pollen. Male flowers of *Catasetum ochraceum* forcibly glue pollen onto the backs of euglossine bees (Romero and Nelson 1986). The experience causes the bees to subsequently avoid male flowers although they are still attracted to female flowers. The stamens of *Portulaca grandiflora* respond to touch by rapidly bending towards the stimulation. This is believed to allow the stamen to push up against the backs of insect visitors, enhancing pollen deposition (Jaffe et al. 1977). For some legumes the weight of the insect "trips" the stamens, forcibly depositing pollen on the body of the visitor (Willmer 2011). All of these mechanisms increase the likelihood that visitors will move pollen instead of merely collecting floral rewards.

There has been disagreement about the potential for plants to restrict their rewards to particular visitors that provide the most reliable pollination services. Many flowers are pollinated by a wide variety of visitors (Waser et al. 1996). Conversely, some flowers communicate using unique odors or combinations of odors that can be sensed only by a limited set of potential players (Raguso 2008). These so-called private channels show a high degree of specificity involving only one visitor species. It is still not clear how common private communication channels are in nature.

Some plants produce nectars that contain secondary chemicals, which are toxic to many visitors, a surprising finding since nectars have historically been viewed as rewards (Adler 2000). Nectars containing secondary compounds may restrict visitation to effective pollinators, alter visitor behavior to make them more efficient pollinators, inhibit microbes, or be caused by some other form of selection or nonadaptive force. Experimental enhancement of alkaloid concentrations in the nectar of *Gelsemium sempervirens* reduced pollen export but had little effect on female function (Adler and Irwin 2005). However, for *Nicotiana attenuata*, the presence of both volatile attractants and nicotine in nectar were required to maximize visitation by pollinators and measures of male and female fitness (Kessler et al. 2008). In a comparison of species of *Nicotiana*, those relying on pollinators for outcrossing had lower concentrations of nicotine in nectar, floral, and leaf tissues than self-compatible species (Adler et al. 2012).

Although some flowers attempt to restrict visitation, other flowers are visited by animals that extract the rewards but fail to contact the anthers or stigma (Sprengel 1793, 1996; Irwin et al. 2010). Nectar robbers most often bite holes in the side of the corolla to reach the nectaries (fig. 7.2). This not only removes nectar rewards without pollinating but also makes the flower less attractive for subsequent visitors that are legitimate pollinators. As a result, nectar robbing was found to depress female fitness in a meta-analysis, particularly for species that required outcrossing (Burkle et al. 2007). Self-compatible plants were less affected, and some even benefited by

FIGURE 7.2 A bee chewing a hole through the calyx of a *Salvia* flower to rob the nectar reward instead of entering through the mouth of the corolla tube. Photo courtesy of Kathy Keatley Garvey.

nectar robbers that also accomplished some level of pollination (e.g., Morris 1996). Although relatively few studies have examined this issue, nectar robbing decreased male function, with some exceptions (Irwin et al. 2010).

Plants have evolved many lines of defense that discourage nectar robbers and other florivores (McCall and Irwin 2006, Irwin et al. 2010). These include physical resistance mechanisms such as thickened calyces (e.g., Roubik 1982) and secondary chemicals in floral tissues and nectar (Adler 2000, McCall and Irwin 2006, Kessler and Halitschke 2009). These traits may have numerous effects on multiple players, so that whether they actually provide defense (increase plant fitness) is not well understood (Irwin et al. 2010). Since nectar robbers compete with legitimate pollinators for floral resources, plant traits that favor the latter may increase plant fitness. For example, bumblebees may act as either robbers or legitimate pollinators; providing artificial flowers with nectar guides increased the likelihood that bees would enter through the mouth of the corolla tube, thereby pollinating the flower (Leonard et al. 2013).

Plants simultaneously interact with many abiotic and biotic agents, all of which can influence selection on floral traits (Irwin et al. 2004, Strauss and Whittall 2006). The same floral traits that attract pollinators may also attract florivores and other herbivores (Brody and Mitchell 1997, Irwin et al. 2003, McCall et al. 2013). Plant defenses that were induced after herbivory have been shown to reduce attractiveness of plants to pollinators and reduce pollinator services (Strauss et al. 1999). Plants can sometimes successfully separate the competing functions of attracting pollinators and repelling herbivores. For example, some *Dalechampia* species have moveable bracts that protect flowers from most insects that are not pollinators, but that allow pollinators to access flowers (Armbruster 1997). *Acacia* trees produce nectars that attract aggressive protective ants as well as pollinators, and the ants can potentially interfere with pollinators (Raine et al. 2002). To prevent this interference, plants may use bribes such as extrafloral nectar to attract ants away from flowers, volatiles to repel ants from flowers, and physical barriers to isolate flowers from ants (Willmer et al. 2009). Nectars that attract pollinators are often rich in sucrose although *Pseudomyrmex* ants that inhabit *Acacia* trees lack the enzymes to digest sucrose and prefer nectars that have other sugars instead (Heil et al. 2005). These nonsucrose nectars are provided by extrafloral nectaries and are not attractive to ant species that are not *Acacia* specialists (Heil et al. 2005). Specialist ants avoid flowers, particularly pollen, and rarely overlap spatially with visitors to flowers (Raine et al. 2002, Nicklen and Wagner 2006).

7.1.4 Plant-pollinator communication and plant speciation

Animals serve as pollen vectors, facilitating mating in many plants. Plants influence this process by offering rewards to flower visitors along with cues that animals perceive and respond to. Plants that provide different rewards and different cues will attract different visitors. This has the potential to cause plant populations to diverge, to become reproductively isolated, and to speciate (Grant 1949, Grant and Grant 1965, Johnson 2006).

The hypothesis that pollinators and the corresponding floral traits that they select can drive plant speciation has been around at least since Darwin (1859), although it remains controversial. The evidence is compelling that flower visitors exert strong selection for particular floral traits and that these traits can enhance rates of pollination. For example, selection for mating success has shaped corolla morphology in *Ipomopsis aggregata* (Campbell et al. 1997) and *Polemonium viscosum* (Galen 1996, 1999). Most pollination biologists agree that pollinators have caused floral traits to diverge, and that particular traits (syndromes) are associated with particular groups of pollinators (Fenster et al. 2004). Plants that are pollinated by animals have radiated and are currently much more speciose than those pollinated by abiotic vectors (Dodd et al. 1999). Both as agents of selection and as agents of gene flow, pollinators will be most effective at causing floral divergence and plant speciation if they are specialized to visit only certain flowers (flower constancy) (Waser 2001). What remains unclear is whether pollination systems are sufficiently specialized to lead to ethological isolation and speciation (Waser et al. 1996, Waser 2001).

7.1.5 Communication and postdispersal pollen fate

Most work on the interactions between plants and pollinators concerns the movement of pollen from anthers to stigmas, a bias that has been reflected in this chapter. However, to be represented in the next generation, a pollen grain must germinate on a stigma, grow a pollen tube that competes with other pollen grains in a race to reach an ovule, and fertilize the ovule. This fertilized embryo then competes with others for maternal resources. During this process, the maternal plant can provide signals that help to direct the growth of the pollen tube and may evaluate the pollen or developing embryo and provision it richly, or not at all.

Interactions between pollen grains on the stigma can affect pollen success. Proteins and lipids on the pollen coat recognize the stigma and facilitate adhesion to this surface (Chapman and Goring 2010, Chae and Lord 2011). Following adhesion, a protein secreted by the stigma interacts with pollen receptor kinases to promote pollen tube growth (Tang et al. 2004). Pollen germination rates increased as the density of pollen grains increased on stigmas of numerous species (Brewbaker and Majumder 1961). This situation involving high pollen grain density has been interpreted as providing increased pollen competition. Females may also delay stigma receptivity, which should increase pollen competition and maternal fitness (Lankinen and Kiboi 2007, Madjidian et al. 2012). Females can also increase competition among potential sires by lengthening the pistil and manipulating its pH (Skogsmyr and Lankinen 2002).

Pollen tubes must penetrate the stigmatic cuticle and grow through the cell wall, a process that requires enzymatic modification of these barriers; both the pollen tube and stigma likely contribute to this process (Chapman and Goring 2010). Pollen grains from different males interfered with one another while those from the same male stimulated each other under some conditions, although the mechanisms responsible have not been determined (Skogsmyr and Lankinen 2002).

When pollen grains germinate they are dependent on their own nutrients but as they grow they become dependent on nutrients supplied by the female style (Herrero and Hormaza 1996). Pollen tubes grew more quickly and persistently in the pistil tube than in an artificial medium (Chae and Lord 2011). Directed polar growth towards the ovule was aided by signaling between the male pollen tube cell and the walls of the transmitting tract within the female pistil. The mechanisms of this signaling appear to be species-specific (Chae and Lord 2011, Takeuchi and Higashiyama 2011). The transmitting tissue within the pistil provides chemical gradients and nutrients that orient the pollen tube and enable its growth (Chapman and Goring 2010). The female target produces D-serine, an unusual amino acid that interacts with glutamate-like receptors in the pollen tubes, activating Ca^{2+} channels that coordinate the growth of pollen tubes (Michard et al. 2011). At close range, proteins secreted by the ovule serve as chemical attractants that further orient the growth of pollen tubes (Okuda et al. 2009). Access by the pollen tube to the ovule through the micropyle is controlled by the female gametophyte, and chemical gradients direct pollen tubes towards unfertilized ovules (Hulskamp et al. 1995, Chapman and Goring 2010, Chae and Lord 2011).

Most of the control of pollen tube growth appears to reside in factors produced by the pistil, although the pollen tube needs to perceive these cues and grow rapidly towards the ovules (Chapman and Goring 2010). These interactions give female plants the opportunity to interact with, and influence, pollen tube growth (Stephenson et al. 2003). In several plant species, pollen tube growth correlated with offspring fitness so that females may be able to sort among various pollen donors (Skogsmyr and Lankinen 2002).

Interactions between pollen and pistil can also result in self-incompatibility reactions that prevent inbreeding. Species-specific reactions reject or destroy self-pollen when it is encountered (Poulter et al. 2010). Pollen grains with the same incompatibility alleles as the stigma will not adhere to the stigma and will not produce growing pollen tubes; nor will most heterospecific (hybrid) pollen (Fujimoto and Nishio 2007). Self-incompatibility has evolved numerous times, and different plant families employ different mechanisms. In the Brassicaceae, if a pollen grain with the same haplotype as the recipient does attach to the stigma, a receptor kinase is activated, leading to a cascade of reactions that block pollen hydration and growth

(Chapman and Goring 2010). In poppies, programmed cell death of the self-pollen follows and involves a Ca^{2+} signaling network (Wheeler et al. 2010).

Self/nonself-recognition is determined by a single locus, the S locus on both the pollen and the stigma. Recognition of self-pollen occurs when S alleles of the pollen and stigma match (Wheeler et al. 2010). The S genes are highly polymorphic, with as many as 60 different alleles possible. In the Brassicaceae, S genes code for proteins associated with the pollen coat; these proteins interact with receptors on the stigma to determine whether pollen is accepted (Takayama and Isogai 2005). In other plant families, S genes code for either low-molecular-weight proteins or ribonucleases that are secreted onto the stigma and recognize self-pollen (Wheeler et al. 2010, McClure et al. 2011). RNase can act as a cytotoxin that degrades self-pollen; this degradation is prevented in compatible interactions. These diverse mechanisms all involve complex signaling between male and female cells that reduce the frequency of self-fertilization and inbreeding.

7.1.6   Selective abortion of ovules

Female plants often produce more ovules than they can mature, allowing them to selectively abort some ovules (Lee 1984). Female plants must first be able to recognize more or less desirable traits in the developing embryos. According to this hypothesis, females can evaluate the potential embryos in terms of their expected fitness later in life and selectively provision those with the best prospects, thereby increasing their probability of success. Many empirical results were consistent with this hypothesis. For example, abortion rates were lower for ovules fertilized by larger pollen (Cruzan 1990), by outcrossed pollen (Montalvo 1992), or for fruits with multiple pollen donors (Marshall 1988). However, in these and other examples there was little ability to evaluate whether abortion was caused by maternal control or by competition among embryos (Marshall and Folsom 1991). Genetic approaches, comparing the representation of alleles in seeds that were provisioned and those that were not, have been relatively uncommon, but results have been consistent with selective embryo abortion (Korbecka et al. 2002). More work is required before the role of selective abortion as a means of mate choice can be evaluated definitively.

Seed abortion may also depend on the expected risks and prospects for success. For example, barberry plants sometimes abort seeds that have been attacked by seed-eating flies, killing the flies (Meyer et al. 2014). When fruits contained two seeds, they were more likely to be aborted than when they contained only a single seed. Seed abortion of the infested seed increased the chances that the second seed in the fruit would survive. When fruits contained only a single seed, nonabortion was the better strategy since not every oviposition puncture (the cue) was associated with fly development and seed predation. Barberry plants thus made conditional decisions that integrated different sources of information (number of seeds, cues of attack) that optimized the production of viable seeds.

7.2 Seed dispersal and communication

Plants spend most of their lives rooted in one place; seeds represent the one best opportunity to move to a different location. Plants must rely on external agents to move seeds from the parent to a favorable location where that individual will live the duration of its life. A majority of plants are dispersed by animals in both temperate and tropical ecosystems (Howe and Smallwood 1982). Dry habitats tend to have fewer plants that are dispersed by animals and more species that are dispersed by wind; plants growing in, or adjacent to, water often have seeds that are dispersed by water.

There are several reasons why seed dispersal from the parent plant to new locations will be advantageous (Howe and Smallwood 1982). (1) Dispersal allows seeds to move away from sites of high density where rates of seed and seedling mortality may be high. According to this escape hypothesis, or Janzen-Connell hypothesis, high seed density near the parent leads to higher mortality caused by density-dependent competition, predation, and disease (Janzen 1970, Connell 1971). (2) Dispersal allows seeds to colonize new habitats that are distant from the parent. (3) Dispersal may facilitate placing seeds into specific situations that are both very favorable for their development and very rare. For example, some plant species germinate and grow well only in ant mounds, and seeds are more likely to colonize ant mounds if they are dispersed by ants rather than by any other means (Culver and Beattie 1980). Empirical work has generally provided evidence supporting these hypothetical benefits of dispersing away from the parent (Harms et al. 2000, Bagchi et al. 2010, Terborgh 2012).

7.2.1 Plants offer rewards for seed dispersal

Seeds that are not dispersed by animals are generally light enough to float on air or water currents. Plants relying on these abiotic means of dispersal do not have to provide rewards to animals, but they also have little ability to influence the fate of their seeds. Seeds dispersed by abiotic forces, especially wind, tend to be smaller than animal-dispersed seeds, meaning that they start life with fewer resources for growth (Hughes et al. 1994).

Some seeds are dispersed by animals without providing rewards; the dispersing animal gains nothing in the process. Adhesive fruits attach to the fur or feathers of mammals and birds and hitchhike from the parent plant to some other location where they are groomed off or fall off (Sorensen 1986). Seeds with hooks, barbs, or sticky substances that attach to animals experience low rates of removal since birds and mammals are not attracted to the seeds or to other associated advertisements. Seeds that are dispersed in this way are unlikely to be deposited in particularly favorable locations although they may be moved greater distances than those that are wind dispersed or carried internally by animals (Sorensen 1986).

Some seeds are collected by granivores that consume them (Vander Wall 2001). In these cases, the reward that the plant offers is the seed itself. Presumably the plant provisions the seed with resources that can be used by the embryo and these resources nourish the granivores instead. Seeds (nuts) that are dispersed in this manner tend to have relatively high concentrations of lipids and carbohydrates. Animals, especially mammals and birds, often collect seeds and move them to one or many caches before eating them (Vander Wall 2001). Caching is particularly common when seeds are produced only at one time and granivores store them for use throughout the year, a behavior made more likely by large, infrequent seed crops (masting). The plant benefits only if the seeds are moved some distance from the parent and then the disperser fails to consume them, either because it forgets their location, it dies, or it consumes other food first, allowing the seeds to germinate. Granivores may have single caches where seeds collected from many foraging trips are aggregated or many caches that are dispersed throughout the landscape, a practice called scatter hoarding. The effectiveness of this dispersal method ranges from essentially 0% to as high as 10% of the seeds produced, depending upon many conditions (Forget 1992).

Seeds that are dispersed by ants sometimes offer a small lipid-rich attachment, termed an elaiosome, which ants collect (Beattie 1985). Generally ants carry these seeds from the parent plant to their nests, remove the elaiosomes, and discard or consume the seeds. The benefits (and costs) of granivory by ants accrue to seeds with or without elaiosomes. Even in cases where the ants are granivores and consume many seeds, they can provide effective dispersal similar to the vertebrate granivores considered above

(Levey and Byrne 1993, Retana et al. 2004). Ants often place seeds in high-quality sites (refuse piles) that are distant from the parent plant, well fertilized, and protected from other seed predators (Heithaus 1981, Giladi 2006).

Many plants produce fleshy fruits that enclose the seeds and are consumed by animals (Snow 1971). Seeds in fleshy fruits tend to be larger than those dispersed by other means (Hughes et al. 1994). Larger, more nutritious fruits, and fruits that are more difficult to process are more likely to be carried longer distances by animals (Schaefer and Ruxton 2011). Production of fruit rewards can be expensive as fruits can be large, and contain significant amounts of water, sugars, antioxidants, chlorophyll, and other valuable components. Vertebrates, insects, and microbes all consume fruit and compete for access (Janzen 1977). Relationships between fruit eaters and plants tend to be more generalized than those between pollinators and plants; almost all plants have numerous seed dispersers and almost all frugivores consume fruits of many species as well as other resources. Unlike the situation involving pollination, there is much less of a premium on "fruit constancy." Nonetheless, in exchange for the same rewards, some fruit consumers provide much better dispersal services to the plant than others. For example, fruits of *Ocotea endresiana* were commonly collected by five bird species in a Costa Rican forest (Wenny and Levey 1998). Four of these birds deposited seeds near the parent tree in random habitats while only bellbirds predictably moved seeds away from the parent and deposited them in canopy gaps beneath their perches. Seedling survival after one year was higher for seeds dispersed by bellbirds because of reduced infection by fungal pathogens in the canopy gaps.

7.2.2 Plants attempt to control seed and fruit consumers

Because of the high variation in dispersal services provided by different seed and fruit consumers, selection favors plants that influence the process of fruit handling and consumption (McKey 1975, Herrera 1982). To be evolutionarily meaningful, an animal-dispersed seed must remain viable and germinate after the fruit or other reward has been consumed. Some seeds (especially large ones) are not ingested but are discarded prior to fruit consumption; these are more likely to survive (Levey 1987).

Many seeds survive passage through the digestive tract of frugivores, and some may benefit from the process either as the result of scarification or because gut passage removes pathogens or chemicals attractive to seed predators (Janzen 1981, Fricke et al. 2013). Retention time in the animal's gut can determine the outcome of the interaction from the point of view of the seed. Both the physical condition of the seeds and their spatial distribution are likely influenced by retention time. Secondary chemicals in fruits influence retention time, by inducing either constipation or diarrhea (Murray et al. 1994, Wahaj et al. 1998, Tewksbury, Levey, et al. 2008). A mistletoe that is parasitic on *Acacia papyrocarpa* in Australia requires stems of 1–6 mm in diameter for germination and establishment (Reid 1991). Seeds are consumed and defecated by *Dicaeum hirundinaceum* birds, but the sticky seeds adhere to the birds' cloaca rather than falling to the ground. Birds use branches of the appropriate diameter to wipe the seeds from their bodies, resulting in 85% of seeds passed by this species being placed on perches favorable for their growth.

Secondary metabolites in fruits can filter the consumers that will ingest them. For example, capsaicin, the chemical responsible for the kick in chili peppers, is unpalatable to would-be mammalian seed predators and protects seeds against fungal pathogens (Tewksbury and Nabhan 2001, Tewksbury, Reagan, et al. 2008). However, effective avian seed dispersers are not deterred by capsaicin.

Fruits rarely supply a complete diet for frugivores. In addition, secondary metabolites in pulp may limit the amount of any one kind of fruit that a consumer can ingest. It has been suggested that this dietary imbalance may force consumers to travel to procure other foods, resulting in more effective dispersal (Izhaki and Safriel 1989). This hypothesis has remained controversial, with limited empirical support (Schaefer and Ruxton 2011).

Fruit ripening is one way in which plants attempt to control the behaviors of fruit consumers. Unripe fruits contain seeds that are not developmentally mature and these fruits provide few rewards to consumers; they are difficult to digest, provide less nutrition than ripe fruits, and contain higher levels of secondary metabolites (Cipollini and Levey 1997a,b, Schaefer et al. 2003). Unripe fruits are also less conspicuous to consumers since the parent plant does not benefit when they are harvested.

7.2.3 Plants communicate with fruit consumers

Plants can influence who consumes their seeds and fruits and when this process occurs. For example, fruit ripening is associated with an increase in rewards to consumers and also in advertisements, making fruits more conspicuous at a time that can be most beneficial to the plant. Unripe fruits tend to be hard, to be less pigmented, to release lower concentrations of volatile aromatic compounds, to be more astringent, and to remain more firmly attached to the parent plant. Animals approaching a fruit receive information about its developmental status based on visual, olfactory, and tactile cues. These traits have the potential to be true (honest) signals because both the plant and dispersers can benefit from the transfer of reliable information.

Fruits and seeds attract visually oriented dispersers with contrasting and bright colors. Most unripe fruits are green, and photosynthesis conducted locally contributes to their growth (Cipollini and Levey 1991). As fruits mature they add other pigments that diminish photosynthetically active light (Willson and Whelan 1990). This causes their contribution to total carbon acquisition to decline as they primarily recycle internal CO_2 at this stage (Aschan and Pfanz 2003). Ripe fruits that are attractive to birds are most commonly red or black while those attractive to mammals may be any bright color including green, yellow, and brown (Janson 1983, Willson and Whelan 1990). Fruits dispersed by diurnal animals often change color as they ripen while those dispersed by nocturnal animals rarely do. Instead, fruits that are attractive to nocturnal consumers often release strong scents (Lomascolo et al. 2010). Nuts and other seeds that are cached are typically not colorful, presumably because they are less perishable than fleshy fruits and selection may favor those that are not discovered after being cached. These patterns suggest that fruits may exploit the innate sensory capabilities of fruit-dispersing animals. However, the evidence for this is inconsistent and different individuals of the same animal species often prefer different-colored fruit (Willson et al. 1990). Many birds show a preference for fruits with high color contrast and detect them more easily (Schmidt et al. 2004, Cazetta et al. 2009).

In many instances fruit color is an honest signal of nutritional value (Schaefer et al. 2004, Valido et al. 2011). Color provides incomplete information about the ripeness of fruit. In a study comparing color with nutritional value of fruits in a Venezuelan forest, the most common and conspicuous fruits were red and black; these colors were highly detectable but were unrelated to nutritional value (Schaefer and Schmidt 2004). Selection should favor all fruits to be as attractive as possible, independent of the nutrition they provide (Cipollini and Levey 1997b). However, less conspicuous fruits provided more information about nutrition; yellow and orange fruits indicated high protein and low tannins and blue fruits indicated high sugar and high tannins (Schaefer and Schmidt 2004). In other cases, the same compounds may impart color as well as nutritional value (Valido et al.

2011). For example, anthocyanins and carotenoids function both as fruit pigments and as valuable antioxidants (Schaefer et al. 2004). Carotenoid pigments covaried with lipids, and these two share biosynthetic pathways (Valido et al. 2011).

Pigments also serve other functions and may have evolved for many reasons, in addition to communicating with dispersers (Willson and Whelan 1990). For example, fruit pigments may affect rates of attack by pathogens. The anthocyanins in blackberries were found to be potent antifungal agents and darker black fruit curtailed fungal growth much more than lighter red fruit (Schaefer and Ruxton 2011). The carotenoids that color fruit are consumed and recycled by frugivores, who cannot manufacture their own carotenoids; these chemicals can then be used by animals for vision, signaling, and immunological functions (Rothschild 1975).

Morphological traits of fruits can also provide information to consumers. As fruits ripen, cell walls become less hard as the result of hydrolysis and increased cellulase activity (Brady 1987). Increased water content can make fruits softer. Fruit hardness was a reliable cue of sugar content, and hardness and odor may be more useful to primate frugivores than is fruit color (Dominy 2004).

Fruit odors are often much more complex cues than fruit color patterns (Schaefer et al. 2004). Odor bouquets are made up of many different compounds that require specific receptors for detection. This makes specificity or private communication possible although most odor cues from fruits can be perceived quite generally. Communication involving volatiles is limited in terms of directionality although it is not limited by light and does not require a direct line of sight. These traits make odor a particularly useful cue for consumers that locate food at night and within dense cover.

Detailed information about fruit odors and their attractiveness to frugivores is lacking at this time (Schaefer and Ruxton 2011). Concentrations of volatile emissions increase by up to three orders of magnitude as fruits ripen and qualitative changes in volatile profiles are also common (Rodriguez et al. 2012). In particular, esters and lactones that provide the floral and fruity scents increase during this process. Ethanol production was associated with sugar content in fruit, and some fruit consumers were attracted to the smell of ethanol (Dominy 2004), although higher concentrations were repellent to many frugivores (Goddard 2008). High ethanol concentrations probably represent a situation in which the yeast *Saccharomyces cerevisiae* has outcompeted other larger vertebrate consumers of fruit (Janzen 1977, Goddard 2008). Insects that consume overripe and rotting fruit commonly use the odors emitted by microbial infections of fruit to find their hosts (Hammons et al. 2009, Rodriguez et al. 2012). Larvae of fruit flies seek out the odors of infected fruits and gain protection from their parasitoids by consuming ethanol (Milan et al. 2012). Insect frugivores that consume fruit but do not disperse seeds are often attracted by the same volatile cues as vertebrates, preferring ripe fruits over immature ones (Rodriguez et al. 2012). Some pathogens also appear to require the volatile emissions of ripe fruit in order to successfully recognize and infest fruit (Rodriguez et al. 2011). Despite the appealing logic that fruit colors, odors, and morphologies provide signals for fruit consumers, we still know relatively little about communication between fruits and animal frugivores.

In summary, plants communicate with visitors to flowers and fruits in an attempt to provide information that results in out-crossing and successful seed dispersal. In some cases, information is the only resource provided; in many cases information accompanies other resources such as nectar and pollen for visitors to flowers and nutritious pulp for visitors to fruit. This communication can greatly affect the fitness of both plants and their visitors (see sections 9.2.1 and 9.2.3) although these interacting organisms usually have very different goals.

8 Microbes and Plant Communication

8.1 Microbes are critical for plant success

Microbes are ubiquitous, and interactions between plants and microbes are enormously important and widespread. Chloroplasts, the organelles responsible for photosynthesis in plants, originated as cyanobacteria that were subsequently engulfed by eukaryotic cells (Keeling 2004). Mitochondria, responsible for respiration in both plants and animals, originated as aerobic bacteria that were engulfed by a prokaryotic organism (Margulis 1971). Because the recognition and incorporation of these microbes occurred so long ago, we have relatively little understanding of the processes that led to these associations, although they play critical roles in allowing life as we know it.

Plants and microbes currently engage in many diverse interactions that require plants to recognize microbes and communicate with them. Some of these interactions benefit the plants and some harm them. The distinction between these two outcomes is often determined by the biotic and abiotic environments in which the specific interactions take place. For example, soybean plants recognized and responded to bacteria that they encountered on the leaf surface (de Roman et al. 2011). These responses provided defense against potentially harmful pathogens although they also made plants less likely to be colonized by putatively beneficial arbuscular mycorrhizal fungi. The environment-dependent costs and benefits of harmful pathogen attack and beneficial mycorrhizal colonization determine the realized net effects of these immune-like behaviors. Because microbes are ubiquitous and difficult to see and study, our understanding of plant-microbe interactions is relatively limited; this is an area of research that is likely to grow rapidly, aided by new technological advances.

In this chapter we will consider how plants recognize and protect themselves against pathogens but encourage relationships with those particular microbes that can increase plant fitness. These later include attracting the microbial enemies of pathogens and herbivores, as well as acquiring mycorrhizal fungi that aid in procuring resources and bacteria that can provide usable nitrogen.

8.2 Plants recognize pathogens

Many viruses, bacteria, fungi, and other microbes feed on living or dead plant cells; these microbes are considered pathogens when they harm plants. Most plant defenses against pathogens require that the potential host plant be able to recognize the pathogen and mount a coordinated defense against it. These recognition processes were considered in section 2.3.3.

Plants are colonized by a nonrandom collection of microbes, and they actively attract or filter potential colonists (Lundberg et al. 2012, Junker and Tholl 2013). Certain microbe species are recruited in preference to the majority of available species, and this sorting is believed to be caused more by differential survival than by dispersal or colonization limitations. For example, *Arabidopsis* flowers produce and emit (E)-β-caryophyllene, which sorts the bacteria that can colonize; plants producing high levels of this volatile reduced the growth of *Pseudomonas syringae* bacteria, but not other microbes, and produced more viable seeds as a result (Huang et al. 2012).

Plants recognize pathogens by reacting to conserved chemical elements that the pathogens possess and the plants lack (Bent and Mackey 2007, see also 2.3.3). In these cases, plants respond to chemical cues of the pathogens or to cues produced by them. Plants also perceive cues emitted by other plant tissues that have been attacked by pathogens. For example, tobacco plants infested with tobacco mosaic virus emitted volatile methyl salicylate (MeSA) (Shulaev et al. 1997). Neighboring plants and unattacked tissues on the infested plant converted this MeSA cue to salicylic acid, causing upregulation of genes associated with defense and providing greater resistance to viral attack. *Arabidopsis* plants also responded to volatiles emitted by rhizobacteria to prime or induce resistance against attack by more damaging pathogens (Farag et al. 2013). Some volatiles emitted by plants that have been attacked by pathogens are directly antimicrobial (e.g., Goodrich-Tanrikulu et al. 1995, Zhang et al. 2006), and others act as signals that prime or induce resistance to microbial attackers (e.g., Yi et al. 2009, Vivas et al. 2012). It remains unclear at this time whether volatile communication between individual plants or among spatially isolated tissues on an individual plant will prove to be common and important in defense against pathogenic microbes.

Microbes have the ability to exploit all plant tissues and products and to communicate with other consumers. As a sugar-rich product, nectar is particularly susceptible to microbial degradation that reduces its attractiveness to pollinators. Similarly, fleshly fruits contain nutrients that make them attractive to vertebrates that might disperse their seeds but also attractive to microbes that consume them without providing these services. Microbes may exclude other consumers by making nectar and fruits unpalatable to other microbes and to potential insect and vertebrate consumers (Janzen 1977, Davis et al. 2013). What we perceive as rotten fruit and fermented nectar represent resources that particular microbes have modified for their own exclusive use. For example, many different microbes colonize ripe fruits, but one yeast, *Saccharmyces cerevisiae*, is particularly tolerant of the high ethanol concentrations that it causes; this species comes to dominate in competition for these resources (Goddard 2008). This same species has also been used for thousands of years to make wine, beer, and bread. Microbes that alter the quality of nectar or fruit are more likely to benefit if they communicate their presence to potential competitors. Visual and chemical cues can potentially alert animals to the presence of microbes (Davis et al. 2013), and even bacteria have the ability to sense other microbes and distinguish self from nonself (Gibbs et al. 2008).

As plants defend themselves against ever-present microbes, those defenses can have indirect and nuanced consequences. For example, plants respond to many biotrophic pathogens by inducing systemic acquired resistance (SAR) mediated by the salicylic acid (SA) pathway (Glazebrook 2005) (see sections 3.3 and 4.3.5). One common outcome of this pathway is a localized hypersensitive response that kills infested plant cells and isolates the pathogens associated with those dead cells. This strategy is effective against biotrophic microbes that feed on living plant tissue but can be counterproductive when facing necrotrophic microbes that prefer dead tissue. In other cases, resistance to some pathogens is driven by signaling involving jasmonic acid (JA) and ethylene; these signals are also associated with resistance to many herbivores (Bostock 2005). The quantity, composition, and timing of the signals produced by the plant is determined by the lifestyle and infection strategy of the invading attacker, and these nuances allow plants to tailor their defenses to particular microbes (Pieterse et al. 2009). However, these various phytohormones may have antagonistic effects that prevent plants from being simultaneously defended against all pathogens and may be manipulated by some attackers. For example, it has been known for some time that SA interferes with the JA pathway and with expression of resistance to herbivores and some pathogens (Doherty et al. 1988, Doares et al. 1995). Induction of SA can depress JA synthesis and responsiveness of plants to JA (Spoel et al. 2003). The result is that plant responses to one microbe often make it more susceptible to other pathogens and herbivores (Bostock 2005, Stout et al. 2006, Thaler et al. 2012).

Responses to microbes interact with other plant processes. For example, low light intensity has a complicated and generally negative effect on SA signaling (Kurepin et al. 2010). Abscisic acid is involved in plant signaling in response to abiotic stresses and it also antagonizes SA-dependent defenses (Yasuda et al. 2008). Other phytohormones that regulate various plant functions simultaneously interact with SA, JA, and ethylene to shape the resulting plant immune responses (Pieterse et al. 2009).

Some microbes produce plant hormones and are able to hijack the host's signaling apparatus for their own benefit. The ploys of the bacterium *Pseudomonas syringae* are particularly well documented (Pieterse et al. 2009). Various chemicals produced by this pathogen have been shown to alter auxin physiology, influence abscisic acid signaling, and act as powerful JA mimics, all of which resulted in suppressed defenses and enhanced disease severity.

8.3 Infested plants attract the microbial enemies of their attackers

Some microbes attack the enemies of plants, and plants sometimes favor these strains and even communicate to attract them. Roots of white cedar that were attacked by herbivorous insects emitted chemicals that probably moved through the groundwater to attract entomophagous nematodes that parasitized these herbivores (Van Tol et al. 2001). Maize roots that were attacked by beetle larvae emitted a volatile sesquiterpene that attracted entomophagous nematodes (Rasmann et al. 2005). Maize varieties that produced this volatile achieved nematode infection rates of their insect pests in the field 5 times as great as those that failed to produce the cue. The combinations of CO_2 and specific volatiles were found to be most effective as cues that attracted entomophagous nematodes (Turlings et al. 2012). Roots of milkweed plants that were infested with beetle herbivores emitted a complex blend of compounds that attracted entomophagous nematodes (Rasmann et al. 2011). In this case, nematodes reduced the survival of beetle larvae so that plants emitting cues perceived by nematodes were as successful as plants with no beetles in terms of production of aboveground biomass.

Volatiles produced by plants attacked by foliar herbivores also communicate with, and otherwise affect, fungi that infest herbivores. For example, volatiles from uninfested cassava normally reduce the production of conidia of mite-pathogenic fungi (Hountondji et al. 2005). However, when cassava plants were infested by mites they produced volatiles that stimulated the production of infectious conidia.

Some microbes have been found that prime plants and induce greater resistance to other pathogens or herbivores, an interaction called induced systemic resistance. Most examples involve soil bacteria that potentiate plant responses against aboveground pathogens and insects (van Loon et al. 1998). Increased sensitivity to jasmonic acid and ethylene, but not salicylic acid, are involved, although the mechanistic details of this phenomenon are not well understood (Zamioudis and Pieterse 2012). Even with a limited understanding of the mechanisms, agriculturalists have had some success manipulating rhizosphere microbes to control plant pests (Vallad and Goodman 2004).

8.4   Plants communicate with mycorrhizal fungi

Roots are the plant organs that absorb water and nutrients from the soil, although roots are not particularly good at these tasks. The capacity of roots to absorb nutrients is enhanced by associations with mycorrhizal fungi because fungal hyphae are finer than plant roots and often extend beyond the limits of the root zone, greatly enlarging the soil volume for absorption (Clarkson 1985). Some authors have argued that mycorrhizae, and not roots, are the chief organs of nutrient acquisition for land plants (Smith and Read 2008). Approximately 80% of plant species are associated with mycorrhizae—either endomycorrhizae whose hyphae penetrate plant cells or ectomycorrhizae that do not penetrate cells but are abundant enough to be comparable in mass to their host roots. Mycorrhizae are particularly good at taking up phosphorus, other minerals such as zinc and copper, and water. In exchange for nutrients, plants supply the mycorrhizal fungi with carbohydrates.

Plant-mycorrhizal relationships may be context dependent and facultative (Smith et al. 2009). Phosphorus-deficient situations make mycorrhizae particularly valuable and cause plants to allow or encourage the relationship by supplying more carbon. In fertilized soils, mycorrhizae have less to contribute and may become parasitic on their hosts. Under these conditions, plants may treat mycorrhizae as they would other pathogenic microbes and restrict their interaction (Brundrett 1991).

There are far fewer species of mycorrhizae than of plants; most mycorrhizae are generalists, capable of associating with many different plant species to a greater or lesser extent (van der Heijden et al. 1998, Smith and Read 2008). Extensive mycorrhizal networks develop that exchange carbon and other nutrients, linking numerous plant individuals and species (Simard et al. 2012). Mycorrhizal networks transfer water between plants along with any other compounds that are water-soluble. These include amino acids, allelochemicals, and informative chemical signals (He et al. 2005, Barto et al. 2011).

Plants may communicate with one another (or eavesdrop) using information transferred via mycorrhizal networks. Some plants sense and respond to cues that indicate the experiences of their neighbors in this manner. For example, uninfested tomato plants that were linked by mycorrhizal networks responded to pathogen infection of connected neighbors; the uninfested plants elevated levels of resistance to pathogenic fungi before being attacked themselves (Song et al. 2010). Since mycorrhizal networks may cover meters, communication through these networks may be possible over greater distances than have been detected in research involving airborne cues. Bean plants that were linked by mycorrhizae to neighbors infested with aphids emitted volatiles that repelled aphids and attracted their parasitoids (Babikova et al. 2013). In this case the bean plants prepared for aphid attack despite no direct contact with these insects.

Colonization of new plant host individuals by mycorrhizae is highly conditional. Plants can signal their location to recently germinated fungal spores (Giovannetti et al. 1996, Akiyama et al. 2005). Plant roots release a sesquiterpene recognized by the mycorrhizal fungi that causes increased hyphal branching, leading to contact with the root and ultimately root penetration. Plants mount an SAR response that involves SA when they perceive most fungi. Similarly, exogenous experimental applications of SA delayed mycorrhizal colonization (Pozo and Azcon-Aguilar 2007). However, the SA responses of plants were attenuated during compatible interactions between roots and mycorrhizae (Liu et al. 2003, Zamioudis and Pieterse 2012). Some mycorrhizal fungi appear to secrete proteins that act as effectors, hijacking plant signaling that would otherwise destroy them (Kloppholz et al. 2011). In addition, once plants recognized mycorrhizae, they redistributed nutrients and structures to accommodate the mycorrhizal partner inside root cells (Genre et al. 2005).

Plants may regulate their interactions with already established mycorrhizal associates by sensing the contributions of each partner and adjusting the quantity of resources that they provide to each in return. A fungal partner that delivered more phosphorus was provided with more carbon from that host plant (Kiers et al. 2011). This allowed plants to reward generous partners and to punish cheaters or less cooperative associates. The converse was also found; mycorrhizal fungi receiving more carbon from a host provided more phosphorus in return. A similar reward system also ensures that trade remains fair between mycorrhizal fungi supplying nitrogen and plants supplying carbon (Fellbaum et al. 2012). Communication between the partners is an essential element of this interaction. This system of rewarding cooperative partners prevents either of the associates from being enslaved by the other.

8.5 Plants communicate with N-fixing bacteria

Unlike most animals, plants can synthesize all of the amino acids that make up their proteins, although plants are often limited by the supply of available nitrogen precursors. Bacteria can convert atmospheric nitrogen into usable ammonium. Most of these N-fixing bacteria live independently in the soil; conversely, plants are not always dependent on bacteria since plant roots independently seek out and acquire these resources when they are available in the soil. However, some plants form symbiotic relationships with N-fixing bacteria, housing the bacteria in specialized structures called nodules and supplying the bacteria with carbohydrates and other nutrients in exchange for biologically available nitrogen. These symbiotic relationships have formed numerous times and involve different plant families and different soil bacteria (Franche et al. 2009).

The relationship between plants and N-fixing bacteria is best known for leguminous plants. In these cases, the relationships are facultative; each partner can exist without the other. When nitrogen becomes limiting, the partners seek each other out by engaging in an intricate dialogue of sophisticated chemical signals (Oldroyd 2013). The plant roots secrete flavonoids and betaines that attract free-living rhizobia bacteria to root hairs. Bacterial genes are then activated and synthesize lipochitin oligosaccharide signal molecules. Legume hosts recognize and respond to specific signal molecules by inducing calcium oscillations in epidermis cells of root hairs. At this point, rhizobia attach to root hairs, induce curling of root hair cells, and become engulfed by the curling plant cells. Enzymes encoded by rhizobia genes cause localized degradation of root cell walls, allowing bacteria to enter the root cells. An infection thread forms as an extension of the plant's plasma membrane that penetrates into the inner cortex of the root until it reaches its target. Cortical cells divide and form a nodule that will house the rapidly dividing rhizobia bacteria. At some point a signal from the plant causes the bacterial cells to stop dividing and to organize into N-fixing "organelles." The nodule forms a vascular system that facilitates rapid exchange of nutrients and nitrogen between the associates. Rhizobia produce ammonium, which the plant converts into less toxic forms before exporting the nitrogen to other organs via the xylem.

As was the case for mycorrhizal interactions, host plants exert some control over the rewards that they provide to rhizobia. Some of the rhizobia associated with soybean nodules were experimentally prevented from supplying their hosts with as much nitrogen by reducing their access to atmospheric N_2 (Kiers et al. 2003). The plants responded to these experimentally less cooperative bacteria by withholding resources so that the reproductive success of these less productive rhizobia was half that of more cooperative control bacteria. That is, plants sense the contributions of individual colonies of established rhizobia and regulate the rewards provided to their microbial associates (Kiers et al. 2003). However, there is no indication that the rhizobia can respond to varying plant rewards. Rather, the rhizobia become completely dependent on the surrounding plant for many aspects of their metabolism, a situation that makes colonies of rhizobia more similar to plant organelles than independent organisms (Oldroyd et al. 2011).

At this point, we know less about interactions between plants and microbes than about plants and macroscopic organisms. Infectious organisms have been greatly underappreciated and understudied although they affect the lives of all larger species (Moran 2002, 2012). However, new techniques are allowing us to identify and manipulate microbes. Molecular tools are required to distinguish microbes because their external morphologies are nondescript. These tools are forcing us to abandon our long-held perspective of what Price (1988) called "Noah's ark ecology," which focused primarily on those large species that were represented on the ark, to the exclusion of microbes, which were seemingly left off.

Many of the interactions that have been discussed in other chapters of this book are probably strongly influenced by microbes. For example, in section 7.1 interactions between plants and floral visitors were considered with reference to nectar availability and quality. Recent evidence suggests that different microbes can affect nectar chemistry, considerably altering the cues that flowers direct towards visitors. The common nectar bacteria found in flowers of *Mimulus aurantiacus* changed the nectar pH and sugars much more than the common yeast (Vannette et al. 2013). These microbes altered the attractiveness of the flowers to hummingbird visitors and affected pollination and seed set. It remains unclear whether this example is unusual since *M. aurantiacus* holds nectar for longer periods of time than many other species, although it seems likely that visitors will generally influence the microbes in nectar and that this can have important consequences (e.g., Aizenberg-Gershtein et al. 2013).

Microbes are important to many essential plant functions, both as facilitators and as parasites, and communication between plants and microbes can influence the outcome of their interactions. In the future, communication between plants and microbes will almost certainly receive intensive study; our view of its importance will also greatly expand.

9 Plant Sensing and Communication as Adaptations

9.1 Plant senses and emission of cues— adaptive traits?

Plants sense cues in their environments and respond conditionally to those cues. We often assume that an individual's responses will increase its fitness, and hence may have evolved for that purpose. However, this is not necessarily the case; responses to cues may produce no measurable fitness benefits or may produce a net cost. They may serve other ecological and physiological functions and may have evolved in another context. In other words, the responses to particular cues that we observe and assume to have been shaped by selection for sensing or communicating may have other important functions that were responsible for their origin or maintenance. While it is difficult to determine why a trait arose or has been maintained, we can deepen our understanding of the functions and distribution of that trait by attempting to ask these evolutionary questions. In this chapter we will consider the sensory abilities and responses of plants as adaptive traits. I will first enumerate the qualities that we will require in order to consider a trait as an adaptation. Next we will evaluate the evidence for considering anthocyanins, shade avoidance responses, and rewardless flowers as adaptations shaped by selection involving sensing and communication.

An adaptive trait performs a function that increases the

relative fitness of the individual that possesses it, compared to individuals possessing other alternative traits (Williams 1966). Adaptive traits are favored by natural selection although not all traits are necessarily the result of selection. For a trait to be favored by natural selection, three conditions must be met: (1) there must be natural variation in the trait, (2) the variation must be heritable, and (3) the trait must be associated with increased fitness relative to individuals with other traits (Endler 1986). These are sometimes called the Darwinian requisites, and we will return to them repeatedly in this chapter.

Some authors have argued that variation exists for essentially all complex plant traits, such as their abilities to sense their environments and communicate what they sense (Rausher 1992). We can observe heritable variation by growing plants under uniform conditions (a common garden) and monitoring the differences in their phenotypes. If the environment does not contribute much to the variation, most of what we observe will result from genetic differences. However, this observation does not imply that the existing variation necessarily includes traits that are "optimal" solutions to any particular design problem. Consider crop plants that have been genetically engineered to respond to caterpillar feeding by releasing insecticidal toxins normally produced by the bacteria *Bacillus thuringiensis* when it sporulates. The fact that no plants sporulate and naturally produce this toxin in response to caterpillars doesn't mean that it isn't beneficial, that is, that it doesn't increase the fitness of individuals possessing such a response. Rather, its absence in nature is more likely caused by the absence of a natural variant of the trait that selection could have acted upon. In other words, there has never been a plant of natural origin that produced the toxin on its own without genetic engineering. We don't have much of a sense about how commonly a lack of genetic variation limits adaptation.

Selection can act only on existing traits that are heritable, the result of genes (or epigenetic factors) that are passed from one generation to the next. Not all traits have a heritable basis. For example, the responsiveness of some animals to cues has been found to depend on their exposure to stimuli during a sensitive period of development (Bolhuis 1991). Chicks of ground-nesting birds will imprint on any large moving object soon after they hatch. The animal behaviorist Konrad Lorenz famously trained goslings to imprint on him, although recognition of cues was not inherited and passed on between generations of geese. While the developmental sensitivity to cues was inherited and subject to selection, the actual cues and specific responses of the birds were not.

Heritability is defined as the ratio of the additive genetic variation to the overall phenotypic variation. Heritability is difficult to measure, and reliable estimates have been made for relatively few of the traits considered in this book. For example, clones of tall goldenrod (*Solidago altissima*) showed variation in their responses to a fly that causes them to form a gall, which houses and feeds the offspring of the fly (Anderson et al. 1989). Instead of forming a gall, some genotypes lacked cues required by ovipositing flies or exhibited a localized hypersensitive response that killed the developing fly larvae. In a common garden 24% of the variation in the abundance of the gallmaker on plants was attributable to differences in plant genotypes (Maddox and Root 1987). Nearly all of this resistance was heritable, estimated by either parent-offspring regression (heritability = 1.12 ± 0.22) or by sib correlation (heritability = 0.92 ± 0.33). Although it is beyond the scope of this book, it is controversial whether heritability is necessarily the best empirical measure of evolutionary potential, the character that we often care about (Hansen et al. 2011).

When heritable traits are consistently associated with increased survival or reproduction, those traits will increase in the population, relative to alternatives. The traits that are most commonly used as examples of natural selection are morphological characteristics such as the beaks of Galapagos finches or the color patterns of moths. However, behavioral traits that could produce the response patterns considered in this book are subject to the same processes.

It can be problematic to consider specific traits as adaptations that resulted from selection for particular functions (Gould and Lewontin 1979). Gould and Lewontin challenged the adaptationist practice of attempting to explain "atomized traits" as optimal solutions designed by natural selection. For one thing, evolution may result from genetic drift, particularly when populations are small. Adaptation by natural selection may be limited by genetic correlations between traits (epistatis) and by traits that cause multiple consequences (pleiotropy).

For example, it has been puzzling why some fruits contain chemicals that are toxic to many potential fruit eaters (Ehrlen and Eriksson 1993). None of the hypotheses that only considered fruit-frugivore interactions provided satisfactory explanations for the existing patterns. One possible explanation is that the fruit characteristics that were being examined were largely shaped by selection for correlated traits that were not the original focus of the investigation. Plants that had high concentrations of toxins in leaves also had relatively high levels of the same toxins in their fruits. The specific traits associated with fruit chemistry may have been driven by selection on leaf traits and animals interacting with leaves and not simply by frugivores interacting with fruits. Similarly, red and purple flowers were found among species of maples that had anthocyanins in leaves while pale flowers were found among species without foliar anthocyanins (Armbruster 2002). It may be impossible to understand floral colors without also considering leaf traits and vice versa. Selection can be slowed by genetic correlations between traits, although we do not know how commonly such correlations actually constrain the evolution of important traits that we may be interested in.

Although we would like to understand the selective factors and evolutionary histories of particular traits, this is an extremely difficult task. Often, the best we can hope to achieve is some understanding of the factors that currently maintain those traits (Endler 1986). Observing that some traits are more successful than others in current environments provides some clues as to why those traits may have persisted or spread but not why they arose initially.

9.1.1 Evaluating fitness consequences

The Darwinian requisites are necessary for a trait to have been shaped by natural selection although satisfying these criteria is not sufficient evidence to conclude that the trait was selected for the function being considered. Imagine a trait that communicates important information about an emitter. That trait may increase the fitness of both the emitter and the receivers. Yet it would be incorrect to conclude that the trait in question necessarily evolved to provide that function rather than for some completely different purpose. For example, a large focal plant changes the quality of light (red:far-red ratio) experienced by its neighbors. Large size communicates information about the sender to neighboring receivers. The signal may cause the neighbors to grow away from the focal plant and may increase the fitness of both the focal plant and its neighbors. Yet it would be silly to conclude that large size evolved to allow the focal plant to communicate with its neighbors (Maynard Smith and Harper 1995).

This example points out that many traits that satisfy the required conditions—variation that occurs naturally, is heritable, and increases relative fitness—may have arisen to fulfill other functions but may have subsequently been molded by selection related to the function being considered. Since communication between individuals of different species requires adaptations by both the sender and receiver, it is particularly difficult to imagine that these interacting systems initially arose simultaneously in both individuals to fulfill such a function. Animal behaviorists who have considered this problem generally assume that the signals first evolved for reasons other than communication and they were subsequently co-opted and changed for their roles as signals (Bradbury and Vehrencamp 1998). Once either the sender or the receiver has a trait that is characteristic of a particular situation, the other party may evolve to recognize or exploit this cue. Many cues are thought to have originated as inadvertent or unavoidable by-products. All organisms inadvertently emit a large number of chemicals that can potentially provide specific detailed information about them to other organisms. For example, plants may recognize pathogens and herbivores based on highly conserved and distinctive chemical cues that these organisms emit (see sections 2.3.3 and 6.1.2). Plants initiate different defensive responses based on the specific chemical signatures of the threats that they perceive.

The net fitness consequences of responding to cues and emitting them will determine whether these behaviors should be considered adaptations (see requirement 3 above). Both the emission and the perception of cues have associated costs and benefits, and an "economic" evaluation of these quantities can allow us to estimate the fitness trade-offs that accrue to the sender and receiver. If signals are not inadvertent by-products of other processes, then their production and emission may be costly for the sender in terms of resources and specialized organs required. Similarly, perception of the cues may be costly to the receiver in terms of receptors and resources.

Energy, estimated as calories, is the most convenient and straightforward currency to measure these costs and benefits. For example, workers studying pollination systems have long recognized that a plant and its visitors have conflicting interests and that each attempts to gain resources (or services) while spending the fewest calories (Heinrich and Raven 1972). A consideration of the caloric gains and losses has provided insights into the net profitability of many interactions. However, while calories are an experimentally tractable currency, they also provide a rather incomplete picture. Sensing and responding to cues have other indirect or ecological costs that are not easily assessed as energy. For example, cues that provide information to beneficial flower visitors may also be perceived by harmful predators and pathogens that attack flowers or developing seeds. Responses to one organism may preclude responses to another; these trade-offs may be thought of as opportunity costs. For example, defending against chewing herbivores may preclude effective defense against some pathogens (Bostock

2005, Stout et al. 2006, Thaler et al. 2012). These costs cannot be measured as calories.

The only meaningful currency to evaluate costs and benefits is lifetime fitness, although this is notoriously difficult to measure. Evolutionary biologists are forced to estimate correlates of fitness—survival at different life stages, number of inflorescences produced, seed production and germination, export of pollen, and so on—and hope that these correlates provide an accurate assessment. The measurements that an investigator chooses as estimates of fitness can influence the conclusions about the magnitude of selection (Geber and Griffen 2003).

Responding to cues in a plant's environment has associated fitness costs and benefits, as does failing to respond or responding inappropriately. Emitting signals can also be associated with fitness consequences. Several features of the signaling process can tend to shift the balance towards a net fitness gain or loss. From the point of view of the receiver, the reliability of the information in the cue makes sensing and responding more likely to provide a benefit (Zahavi and Zahavi 1997). Selection will favor only those individuals that respond to reliable (honest) cues. This generalization can have exceptions if the costs of responding to a dishonest cue are small compared to the potential benefits. One interesting exception involves orchids that entice male insects to "copulate" with them, although they offer no actual rewards to the insects. In these cases of pseudocopulation, the insects pay relatively small costs by being fooled. Many of the male insects that visit deceptive orchids show more interest in copulating with orchid flowers when actual females of their own species have not yet emerged (Willmer 2011). When given the choice of mating with conspecific females, the orchid flowers are no longer attractive. This example of rewardless flowers is considered in more detail later in the chapter (section 9.2.3).

For many of the examples of communication involving plants, the participants have overlapping interests, and both can benefit by transmitting reliable information. This will be the case for communication between spatially separated parts of the same individual plant, between closely related kin, and between some obligately symbiotic partners. However, in many possible scenarios involving communication between different individuals, the individual sending the information may be selected to emit an unreliable, dishonest cue (Schenk and Seabloom 2010). Researchers studying animal communication have pondered this problem and now conclude that receivers will be most likely to respond to cues that are costly for the sender to emit (Zahavi 1975, Zahavi and Zahavi 1997). Cues that have some inherent cost to the sender—energy for their production, associated risk of predation, etc.—will be difficult for the sender to fake. Selection will favor receivers that pay attention only to these expensive, and therefore reliable, cues. This conclusion has been supported by theoretical and empirical results in a variety of animal systems (Bradbury and Vehrencamp 1998, Searcy and Nowicki 2005). This is a somewhat counterintuitive argument since senders will otherwise be selected to minimize the costs associated with emitting cues.

Many cues are known to serve multiple functions for the plants that produce them, making it difficult to identify the forces of selection that have shaped them. For example, flower color has conventionally been assumed to be the result of selection to attract pollinators. However, careful analyses have revealed that flower colors may have been shaped by pleiotropic effects of alleles that influence multiple traits (Rausher 2008). Anthocyanins are the pigments that are responsible for red, violet, and blue colors in flowers and fruits. The enzymes that are involved in the production of anthocyanins are also required to synthesize other important flavonoids. In addition to providing color, anthocyanins function as antioxidants that protect against oxidative stress, as sunscreens that protect against damaging radiation, as defenses against herbivores, and as a means of increasing tolerance to cold, heat, and drought (Strauss and Whittall 2006, Schaefer and Ruxton 2011). Anthocyanins influenced correlates of plant fitness that were independent of pollinators and even occurred developmentally before flowers were produced in several species (Strauss and Whittall 2006). Selection on the alleles for anthocyanins may have favored other plant functions even at the expense of traits that would have optimized pollinator attraction (Rausher 2008).

Natural selection can operate on traits that initially evolved to fulfill other functions but can be subsequently shaped as cues and signals. From the sender's point of view, signals that are already produced to serve another function will often be less expensive to produce. Similarly, signals that the receiver is preadapted to sense may be most likely to be selected for communication. There are many examples that are consistent with this hypothesis, although no rigorous tests have been conducted. For example, many pollinators are innately attracted to large and contrasting displays or to particular colors, and these sensory biases can be exploited by flowers (Naug and Arathi 2007, Raine and Chittka 2007). If receivers have sensory biases, they can be tricked by exaggerated signals that have not been common in their evolutionary history (Naug and Arathi 2007). For example, ant-dispersed seeds attract specific seed-collecting ants with oleic acid, a highly conserved compound that is used more generally by ants as a chemical mediator of behavior (Pfeiffer et al. 2010). This compound is highly attractive to ants; therefore, plants that provide nutritional rewards and also less common cheaters that provide fewer rewards both use oleic acid to attract ants that remove their seeds.

9.1.2 Net fitness effects of sensing and communication

With fitness as a currency, evolutionary biologists can categorize interactions between interacting individuals and species based on the net effects of the exchange of information as signals or cues (see fig. 9.1).

Natural selection will favor individuals that produce a cue only when doing so increases the signaler's chances of surviving and reproducing. In some instances production of a signal is also associated with an increase in the receiver's fitness; these mutually beneficial consequences result from so-called true communication. Signals produced by plants that attract and reward effective pollinators can increase the fitness of both. For example, floral cues produced by *Polemonium viscosum* affected the behavior of bumblebee visitors, and this behavior increased the plant's fitness (Galen 1996, Galen and Geib 2007). Similarly, floral cues allow bees and other pollinators to forage more profitably and increase their fitness, although these effects on flower visitors are not as well-known as the beneficial effects of pollinators for plants. For example, visual and odor cues from pollen of novel plant species determined how readily *Osmia* bees collected it (Williams 2003). The ability of bees to locate and harvest novel pollens affected their larval survival, development time, and growth rates. Traits associated with nectar quality have also been found to affect larval performance of bees that visit flowers (Burkle and Irwin 2009).

In some instances, cues that benefit a sender can have minimal fitness consequences on the receiver (manipulation in fig. 9.1) or can cause the receiver to experience a reduction in fitness (deceit in fig. 9.1). As discussed above, orchids that attract insects by producing cues that fool the visitors into copulating with a flower increase the orchid's fitness when pollination is achieved (Willmer 2011). Insects that copulate with flowers rather than fertilizing conspecifics are deceived and may be expected to experience a fitness cost; however, most instances of pseudocopulation occur before females become available and are examples of manipulation of the insects with little fitness costs.

		+	0	-
Fitness of Sender	+	**True communication** *Flowers and bees*	**Manipulation** *Orchids and early bees*	**Deceit** *Orchids and later bees*
	0	**Eavesdropping** *Sagebrush and tobacco*	**Ignoring** *Sagebrush and lupine*	*Sagebrush and tobacco w/ frost*
	-	**Exploitation** *Nectar robbers* *Herbivores*		**Spite**

FIGURE 9.1 Net fitness outcomes of the exchange of information for the sender and the receiver. In many cases, an outcome has a specific name, and these are shown in bold. When applicable, an example of the interaction is shown in italics. Based on Wiley 1983.

Receivers are also under selective pressure to sense cues in their environments and respond in ways that increase their fitness. When receivers intercept cues that were not "intended" for them but use this information to increase fitness, we call this eavesdropping (fig. 9.1). A response by an eavesdropping receiver often has no measurable fitness consequences for the sender. When sagebrush is attacked by chewing herbivores it emits cues that cause neighboring wild tobacco plants to become more resistant to generalist herbivores (Karban and Maron 2002). This often increases the lifetime production of seeds for wild tobacco plants that respond compared to plants that are prevented from receiving the volatile cue. However, under certain circumstances (e.g., a frost), responding to these cues reduces the fitness of the eavesdropping tobacco individuals. The tobacco is small and ephemeral compared to the sagebrush so that its success or failure has little measurable effect on the much larger sagebrush. Other species that were neighbors of damaged sagebrush did not respond in any measurable way, and ignoring the volatile cues presumably had no fitness consequences for either plant (Karban et al. 2004).

A receiver may exploit cues produced by a sender for another purpose, and this may increase the receiver's fitness but decrease the fitness of the sender (exploitation in fig. 9.1). Nectar-thieving ants use the attractive cues produced by flowers to find and consume nectar without providing pollination services; this makes the flowers less attractive to legitimate pollinators and depresses plant fitness (Galen and Geib 2007). Herbivores also exploit the cues emitted by damaged plants, a response that presumably benefits the herbivores and harms the plants (Bruce et al. 2005).

In some instances, communication may harm both the sender and receiver, an outcome referred to as spite. The conditions that favor spiteful communication require individuals to be able to recognize negative relatedness and are poorly studied and probably uncommon (Gardner and West 2006).

9.1.3 Experimental approaches to study adaptation

A powerful tool to evaluate the fitness contribution of a particular trait is to experimentally modify the trait or, alternatively, modify the environment in which the trait is found. If the environment is perturbed and the trait distribution responds in a consistent and predicted direction, this manipulative experiment provides cause-and-effect evidence for selection (Endler 1986). For example, certain floral traits in *Penstamon* flowers were hypothesized to be adaptations that favored visitation by hummingbirds, and others were hypothesized to favor bees (Castellanos et al. 2004). To test this hypothesis, these researchers surgically modified flowers with traits thought to be adapted for visits by bees to make them look like congeners with traits presumed to be adapted for visits by birds. These experimental treatments made bees less likely, and hummingbirds more likely, to transfer pollen although not all of the traits responded as hypothesized.

Genetic tools have given researchers new and enhanced abilities to manipulate traits and examine potential adaptations. Flowers of wild tobacco attract both moth and hummingbird visitors by emitting benzyl acetate, but repel these same visitors with nicotine in their nectar (Kessler et al. 2008). By experimentally silencing the expression of benzyl acetate and nicotine, the functions of these two cues were clarified. The presence of nicotine reduced the time visitors spent at flowers of any individual plant but increased the number of flowering plants that were visited. The combination of both floral compounds increased the number of capsules matured (a measure of female fitness) and seeds sired (a measure of male fitness) relative to plants missing either compound.

Experimental manipulations provide strong inference because the experimenter changes only a single factor and compares that manipulation to a control. We can then confidently conclude that any resulting changes have been caused by the experimental manipulation. Unfortunately the interpretation of selection experiments is not always this straightforward. For one thing, manipulations almost always change more than a single phenotypic trait, be they surgical modifications to morphological traits or genetic changes to biochemical pathways (Dowell et al. 2010). For example, knockouts of genes regulating circadian clocks also affected production of compounds involved in defense against herbivores (Kerwin et al. 2011). A similar network of regulatory connections has recently been found for genes in the jasmonate pathway used to perceive and respond to herbivores (D. Kliebenstein, personal communication). These results suggest that genetic techniques that attempt to manipulate single plant functions will often affect other functions as well.

In addition, experimental manipulations often create phenotypes that do not occur in nature. These may contain traits that are outside of the normal range of variation that natural selection has to work on. Natural selection frequently acts on suites of traits that are coordinated so that artificial selection that creates unnatural combinations may produce misleading results about the potential of natural selection (Campbell et al. 1994, Herrera 2001). While it is important to be aware of these limitations, selection experiments remain a powerful means of learning about causal relationships involving adaptations.

9.1.4 Comparative approaches using phylogenetic models to study adaptation

The research discussed above evaluated the fitness consequences of sensing and communicating for the sender and the receiver of signals or cues. This can provide insights into the forces that currently shape communication systems. We may make the uniformitarian assumption that current selective forces are similar to those in the past that have shaped the adaptations we now see. In some instances where phylogenetic models of the evolution of particular taxa are available, a comparative approach can provide clues about the evolutionary origins of traits, including signaling. This approach examines the correlation between different environmental conditions (presumed selective forces) and different traits (adaptations).

If particular traits are generally associated with particular conditions, this correlation could have been caused by selection that produced a similar solution multiple times in evolutionary history (termed repeated convergent evolution or homoplasy) or it may represent a situation in which species that share traits also share common ancestors (fig. 9.2). In this latter

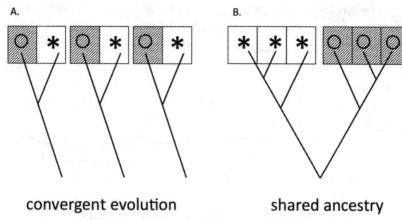

convergent evolution   **shared ancestry**

FIGURE 9.2 Two phylogenetic models showing the evolutionary relationships of six extant species represented by the tips at the tops of the trees. The species have two different traits indicated by the * and the o. The species occupy two different environments indicated by the open or shaded square boxes. In both models, the * trait is associated with the open environment. In model A, the association between the * trait and the open environments evolved three independent times since the three lineages do not share a common ancestor. We might infer that the * trait is favored in the open environment and the alternative o trait is favored in the shaded environment from this model of convergent evolution. In model B, the association between the * trait and the open environment may have evolved only one independent time since the three species with the * trait share a common ancestor and the three species with the alternative o trait also share a common ancestor. This pattern provides less support for the hypothesis that the * trait has been favored in the open environment.

case (fig. 9.2B), the transition to the new trait will likely have occurred only a single time and offers less convincing evidence that the trait is well adapted to those conditions. If a trait is consistently associated with a particular environment and species exhibiting that trait do not share a common ancestor (fig. 9.2A), the trait may have evolved multiple times from convergent selection pressures. This scenario provides stronger evidence for adaptation, although the correlation could also have been favored by selection for a correlated trait or because the trait in question provided other functions that are not being considered. A phylogenetic analysis that includes information about the evolutionary history of the species can distinguish between the two causal hypotheses (shared ancestors or shared environments). It can also provide historical inference about correlated traits—did one precede the other in a majority of cases? A historical perspective can also provide insights into the constraints on evolution and other macroevolutionary patterns.

An example of this approach using comparative methods to gain insights into evolutionary history is the work of Scott Armbruster and colleagues, who studied tropical vines and shrubs in the genus *Dalechampia* (Armbruster et al. 1997, Armbruster et al. 2009). These plants are unusual in offering terpenoid resins as rewards to pollinators. In this case, resins are produced by flowers and collected by several different groups of bee visitors who use them to build their nests. Terpenoid resins are used widely by many different plant taxa as defenses against herbivores. A phylogenetic analysis revealed that more primitive species of *Dalechampia* did not provide resins to flower visitors and were pollinated by bees collecting more usual floral rewards (fig. 9.3A). Early diverging *Dalechampia* species produced resin and used it as a defense of reproductive structures against herbivores and microbes. Resin production was therefore found to have preceded its role as a reward for flower visitors. A few species have more recently evolved resin production by leaves and other vegetative structures, where it functions as a defense of these tissues against herbivores (fig. 9.3B). By mapping the occurrence of these traits onto a phylogeny, it was possible to reconstruct a parsimonious model of the historical origin of these traits.

In another example, *Acacia* species can be categorized as providing high rewards (housing and food) for associated ants or low rewards (Heil et al. 2009). High-rewarding *Acacia* species recognize ants that can provide high levels of defense and provide additional food when these ants are present; low-rewarding species do not show this response. A phylogenetic analysis revealed that high-rewarding plants with obligate ant attendants were ancestral; low-rewarding plants and parasitic exploiter ants without mutualist ancestors have entered the system more recently. Furthermore, mapping these changes onto a phylogeny lent support to the hypothesis that sensing and feedback that depends on the services provided by each partner can stabilize mutualisms.

9.1.5   Genetic correlations and syndromes

Different plant traits may be correlated so that particular suites of traits often occur together. Genetic correlations among different traits may constrain the evolution of trait combinations if particular traits are linked and do not assort independently for either genetic or functional reasons (Geber and Griffen 2003). For example, multiple traits of *Oenothera biennis* were found to be genetically correlated so that particular traits that made plants more susceptible to herbivory were found to be favored by selection because they also allowed plants to increase in size (Johnson et al. 2009).

If traits of flowers that communicate to floral visitors are clustered

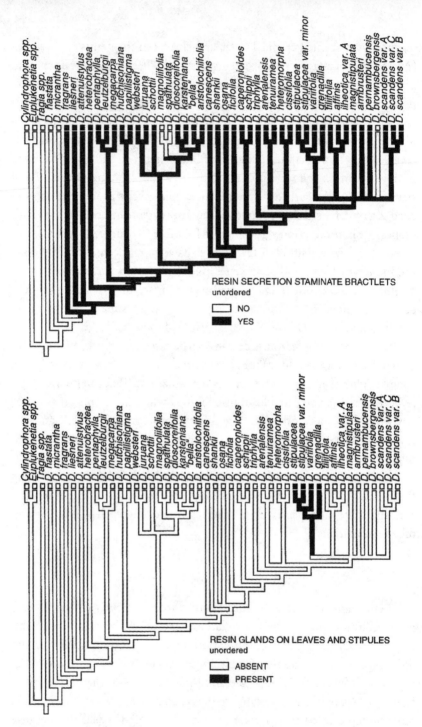

FIGURE 9.3 The distribution and evolutionary history of resin secretion in species of *Dalechampia* spp. inferred from phylogenetic analyses. The three species on the far left are sister taxa that do not secrete resin. A. Resin secretion from bracts. The two most primitive *Dalechampia* spp. also do not secrete resin although most of the other species do; once resin secretion evolved, it was only lost two times. B. Resin glands on leaves are much less widespread and evolved only a single time. Figures from Armbruster (1997).

into pollination syndromes, this clustering may indicate that functionally important traits are constrained by genetic correlations. As mentioned above, genetic correlations could be caused by a single gene locus affecting multiple traits (pleiotropy) or by traits that do not assort independently (linkage disequilibrium). The idea that correlated traits have been shaped by selection by effective, and often specialized, pollinators has been long-standing among pollination biologists (Sprengel 1793, Darwin 1862, Faegri and van der Pijl 1979). More recently, it has become clear that flowers of many temperate species attract diverse visitors and that selection may favor plants that are visited by generalists rather than a limited subset of specialist pollinators (Waser et al. 1996). This has stimulated quantitative assessments of floral syndromes that have supported the earlier anecdotal characterizations. For example, a multivariate ordination analysis of floral traits of 49 *Penstamon* species revealed two distinct clusters represented by species that were pollinated by hummingbirds and those that were pollinated by bees (Wilson et al. 2004). As expected, flower color was important in distinguishing these groups, and other floral traits such as those describing the morphologies of the corollas and anthers were correlated with color and useful as discriminators. This and similar analyses of other plant taxa (e.g., Varassin et al. 2001, Jurgens 2006, Tripp and Janos 2008, Marten-Rodriguez et al. 2009) lend support to the hypothesis that guilds of flower visitors exert similar selective pressures that generate correlated suites of floral traits.

The pattern that has emerged from examining selection on floral traits in the field is one in which different floral traits are phenotypically and genetically correlated (Campbell et al. 1994, Herrera 2001). These clustered traits were termed correlation pleiades (Berg 1960), although some analyses suggest that their frequent co-occurrence is not the result of genetic constraints that limit independent assortment or selection. For example, the floral traits of radish flowers showed strong phenotypic and genotypic correlations (Conner and Sterling 1995). However, artificial selection regimes lasting 5 or 6 generations were sufficient to separate the correlated traits, making genetic constraints unlikely as the force that has been binding them together (Conner 2006). Many members of the Brassicaceae exhibit suites of correlated traits similar to those exhibited by radishes; the exceptions suggest that genetic correlations have been overcome by natural selection as well.

The existence of syndromes has been controversial and their consequences have been largely unexplored. At this point, several conclusions have been reached. (1) There is substantial evidence for the existence of suites of correlated traits that are involved in communication, particularly between flowers and pollinators. Suites of traits often occur together and probably function in an integrated manner. (2) Selection can act on combinations of these traits. Evidence exists for genetic constraints that keep these traits clustered together although selection also appears to be capable of acting on the traits independently. (3) It is important to recognize that the traits can be phenotypically correlated whether or not the genetic architecture forces this correlation.

Recognizing the existence of correlated traits is important because we will want to consider the entire group in order to understand any single trait (Wolf et al. 2007, Sih and Bell 2008). For example, plants that are more perceptive of environmental conditions (the first step in fig. 1.1) may also be favored to be more responsive (the last step in fig. 1.1) because they possess more useful information. Theory has been developed by workers studying animal behaviors suggesting that once an individual possesses a trait that is valuable and not particularly plastic, it may be favored to match this with other traits that function well with this valuable trait (Wolf et al. 2007, Sih and Bell 2008).

Although plant biologists have spent considerable effort determining the existence of pollination syndromes (Faegri and van der Pijl 1979, Waser et al. 1996, Fenster et al. 2004), many questions about correlated "communication syndromes" remain. How constrained are these correlated traits, and why? Do other communication syndromes exist? For example, is there a correlation between a plant's ability to perceive its environment and its likelihood of responding? Is there a similar correlation between a plant's ability to learn and its likelihood of being responsive?

9.2 Case studies of adaptations

In the following sections I will describe three well-developed case studies to illustrate how we have attempted to unravel the selective forces that have shaped plant perception and communication. In the first example, anthocyanin production, we understand the genetic and physiological mechanisms responsible for the cues and some of their fitness consequences. In the second case, shade avoidance responses, we know a little about the genetics of the plant response and more about its fitness consequences. Finally, in the case of rewardless flowers, we know little about the evolutionary mechanisms but have begun to understand the selective forces that allow this counterintuitive trait to persist.

Anthocyanins are pigments found in vacuoles of any tissues of higher plants. They are responsible for the red, purple, and blue colors in flowers and fruits. In these reproductive tissues, they attract visitors that pollinate and disperse seeds. Since they also have many other functions (Shirley 1996) (see discussion above in section 9.1), can they be considered as adaptations for communicating?

For traits to be considered as adaptations there must be natural variation, that variation must be heritable, and the traits must be associated with increases in fitness in the selective environments where they are found. The diversity of floral colors that are found within and among species suggests that variation has existed for natural selection to act on. In at least some instances, this variation is heritable. For example, there are four morphs for petal colors in *Raphanus sativus*, controlled by two independently segregating loci (Irwin et al. 2003). Anthocyanins produce pink petals (A), the absence of anthocyanins results in white petals (a), and pink is dominant to white. Carotenoid pigments produce yellow petals (c), the absence of carotenoids results in white (C), and white is dominant to yellow. Plants with yellow petals are recessive at both loci (aacc), pink dominant at both loci (A-C-), white dominant at the carotenoid locus (aaC-), and bronze dominant at the anthocyanin locus (A-cc).

A more interesting and difficult question involves understanding the selective forces that have shaped flower color. Placing flower color onto phylogenetic models of several different taxa indicated that blue was generally the ancestral state for these species (Rausher 2008). Transitions from blue to red and from highly pigmented to less pigmented (white or light colors) were common and represent more recent evolutionary events. Transitions from red to blue or from less to more pigmented were far less common. Most of the anthocyanins that produce floral colors have a similar basic structure but differ in the number of hydroxyl groups that they contain. Evolutionary transitions from blue to red flowers involved inactivating branches of the anthocyanin pathway, and these were easier to accomplish than the reverse, which involved gaining new functions.

A definitive demonstration that floral anthocyanins are adaptations to attract pollinators requires (1) that changes in anthocyanins and resulting floral color were caused by selection and (2) that pollinators were the agents of selection (Rausher 2008). A review of studies examining the evidence for selection as the cause of changes in floral anthocyanins found support for this hypothesis in 18 of 21 cases (Rausher 2008). Flower color was associated with fitness measures in comparisons involving closely related species, different populations, and different individuals within populations. The evidence strongly supports the hypothesis that flower color is under selection in many natural populations.

It makes intuitive sense that pollinators are the agents of selection for flower color, and this argument has been made repeatedly (Sprengel 1793, Darwin 1862, Fenster et al. 2004). However, anthocyanins are involved in other important plant functions (Strauss and Whittall 2006) (see sections 7.2.3 and 9.1.1). Therefore, it is possible that selection for anthocyanins has occurred but that other agents are also involved, or maybe that other functions are more important drivers than pollinators. One way to evaluate whether pollinators are driving selection on floral colors is to compare the success of progeny from flowers of different colors when pollinators are present and when pollinators are absent and the flowers are hand-pollinated in proportion to the naturally occurring frequencies of color morphs (Waser and Price 1981, Irwin and Strauss 2005). In both of these studies, flower visitors caused progeny to differ in color frequencies compared to the control with only hand pollinations. A similar experiment was conducted comparing the success of red and white flowers that occurred naturally with that of flowers that were painted those colors (Melendez-Ackerman and Campbell 1998). Red flowers were more frequently visited by hummingbirds and were better represented in the generation of progeny than white flowers. These studies provide evidence that pollinators have been the agents of selection in these systems. It is unclear whether this evidence warrants the conclusion that attracting pollinators is generally responsible for selection on anthocyanins and floral color. In no case is there evidence that all of the conditions listed above have been fulfilled in the same species. Some cautious authors are withholding judgment until we have examples of species providing both evidence for selection on flower color and evidence that pollinators are the agents of that selection (e.g., Rausher 2008).

Flower color is probably the oldest and best-studied example of a plant trait that has been considered as an adaptation for communication. Considerable information is available regarding the physiology and genetics of anthocyanins, their fitness consequences, and the phylogenetic patterns of their occurrence. Our inability to draw definitive conclusions even in this case indicates the complexity of questions about adaptation.

9.2.2 Shade-avoidance responses

Plants grow in heterogeneous environments, and they experience sunny and shady patches in close proximity (Smith 1982). The fitness of many plants is limited by their access to light; they are more successful in sunny patches than in shady ones. Phytochrome receptors allow plants to sense their light environments and to respond with a variety of physiological and morphological adjustments (table 5.1). For these sensing abilities and plastic responses to be considered as adaptations, the three requisites described above must be satisfied.

First, there must be natural variation in the sensing and response systems. At a gross level, this requirement is trivial since not all species show the same abilities. However, for microevolution to act there must be variation among individuals or populations within particular species. This assumption has been examined for several species. Natural populations of *Impatiens capensis* and *Geranium carolinianum* differed in their responsiveness to cues of shading (Donohue et al. 2001, Bell and Galloway 2008); families from open environments where shade usually came from similar-sized neighbors were more plastic than those from closed environments where shade usually came from overhead canopies.

Second, variation must be heritable. For *I. capensis*, genetic variation was found in both the shade avoidance traits themselves and in their ability to respond to environmental cues (Donohue et al. 2000). There were limits to the existing genetic variation in these natural populations so that plastic responses to environmental cues were not sufficient to allow plants to achieve the optimal phenotypes in some new environments (Donohue et al. 2001).

Third, plants must exhibit phenotypes that are associated with improved fitness in each environment. Individuals of *Impatiens capensis* that were exposed to the light cues indicative of shading by neighbors produced longer stems and were more fit in environments with high densities of neighbors (Dudley and Schmitt 1996). Individuals exposed to light cues indicative of few neighbors were less elongated and experienced higher fitness in environments with low densities of neighbors. Stem elongation comes at the expense of reduced allocation to roots and sturdy stems (Huber et al. 2004). These allocation costs made shade-avoiding plants more susceptible to drought stress in dry microsites.

Plants are more likely to respond to environmental cues if those cues are reliable in predicting the conditions that the plant will experience (Tufto

2000). There is a lag from the time when the cue is sensed until the trait is expressed in responding tissues. In the case of the shade avoidance response of *I. capensis*, young plants responded to light cues produced by the local density of other seedlings. These turned out to be imperfect cues of the selection on the shade avoidance traits that the developing plants would actually experience in nature (Huber et al. 2004). This unreliability was caused by variability among microsites producing strong effects that were independent of seedling density; the shade avoidance response was more fit in sites with adequate water availability than in dry sites.

There may also be costs associated with growing and maintaining the machinery necessary to sense and respond to environmental cues (DeWitt et al. 1998). Plants that produce the same phenotypes without plastic development do not accrue these additional costs (Van Tienderen 1991). Shade avoidance responses in *Arabidopsis* showed these costs of plasticity, which were detectable in both low- and high-light environments (Weinig et al. 2006). Costs were also detectable for internode elongation in *Geranium carolinianum* but only in high-light environments (Bell and Galloway 2008). In general, costs associated with the ability to respond to light have been small and not frequently detected.

The evolution of plastic responses to cues may also be strongly affected by genetic correlations with other traits (DeWitt et al. 1998). Expression of the shade avoidance response was negatively correlated with defense against herbivores (Kurishige and Agrawal 2005, Moreno et al. 2009, Agrawal et al. 2012). The mechanisms responsible are not completely known although the negative correlation was related to a reduced sensitivity to jasmonate signaling in plants that were experiencing elevated far-red cues (Moreno et al. 2009, Ballare et al. 2012). The negative correlation did not require morphological responses or a diversion of the resources that accompany the shade avoidance response. Jasmonate biosynthesis has also been found to be sensitive to light, which can then affect various defensive products of the jasmonate response (Radhika et al. 2010). The shade avoidance response is exemplary because we understand its genetic basis, its fitness consequences, and some of its costs.

9.2.3 Flowers that offer no rewards

Flower visitors pollinate flowers as an accidental consequence of collecting the rewards that flowers offer, most commonly nectar and pollen (section 7.1). Deceptive flowers advertise that they offer rewards to visitors without actually providing any. Rewardless flowers are found in at least 32 families and have evolved in all major groups of angiosperms, although they are particularly common among orchids (Renner 2006). These species are always rewardless, although individuals of almost all species are facultatively rewardless at times. In other words, individuals of all species will have their rewards depleted temporarily. Rewardless flowers may resemble other flowers that do provide nectar and pollen rewards or they may mimic other resources that the visitors find attractive, such as oviposition sites and receptive females offering mating opportunities for males. It stands to reason that visitors that perceive that flowers are not providing rewards will choose not to visit those flowers, selecting against deceptive and rewardless flowers. Therefore, the commonness of flowers that fail to provide rewards is counterintuitive and demands an evolutionary explanation. Can rewardless flowers be considered adaptations, and why hasn't selection on the part of discriminating visitors eliminated them?

Once again, a useful approach to this problem is to consider whether there is variation in rewardlessness and cues that rewardless flowers offer, whether that variation is heritable, and whether rewardless flowers are associated with fitness benefits that outweigh their costs. Evolution of these traits should not normally be limited by variation. There are numerous examples of permanently rewardless flowers involving taxa that are spread throughout the phylogeny but are embedded within clades of otherwise rewarding flowers (Renner 2006). This suggests that rewardless flowers have evolved independently countless times. In at least some instances this variation appears to be heritable; for example, some populations consistently offer nectar rewards while other populations do not (Brown and Kodric-Brown 1979, Teschner 1980 in Renner 2006). However, the existence of genetic versus environmental variation has not been established in many cases (Mitchell 2004). For example, *Ipomopsis aggregata* is a species known to vary in nectar production although only a small proportion of the variation was estimated to be heritable (Campbell 1996).

Since rewards for visitors are expensive for plants to produce, populations that provide rewards should always be susceptible to invasion by individuals that offer none, unless this selection is opposed by other factors (Bell 1986, Thakar et al. 2003). For example, plants with flowers that lack nectar save on the costs of producing this reward (Ackerman 1986). However, several estimates of the small costs of nectar production have cast doubt on this savings as an important selective agent (Harder and Barrett 1992, Jersakova and Johnson 2006).

An alternative explanation emphasizing benefits rather than costs posits that rewardless flowers have higher fitness because they are more likely to be outcrossed by visitors (Johnson and Nilsson 1999, Jersakova and Johnson 2006). Flower visitors may leave rewardless flowers more quickly and travel greater distances away from patches of rewardless flowers; if true, both of these tendencies could increase outcrossing rates for rewardless flowers relative to flowers offering rewards that might keep visitors foraging locally on an individual plant. This hypothesis has been tested experimentally by adding nectar rewards to otherwise rewardless flowers. The results of these experiments have been mixed and interpretation may hinge on the costs of selfing (Bailey et al. 2007). Adding rewards to nectarless flowers changed bee behavior but did not increase measures of plant fitness for one orchid species (Smithson 2002). Adding nectar to the spurs of another rewardless orchid increased the time that flies spent on each flower and decreased rates of cross-pollination (Jersakova and Johnson 2006). Flies were more likely to move long distances after visiting a rewardless flower, and self-pollinated flowers set fewer seeds than cross-pollinated flowers.

Flowers that provide no rewards to visitors may be successful in situations where they mimic the signals that visitors use to locate actual rewards. For example, rewardless orchids attracted visitors by mimicking the visual cues of their neighbors (Peter and Johnson 2008). Experimentally altering the visual signals using a UV-absorbing sunscreen decreased visits by bees. Proximity to rewarding models was critical and affected visitation rates by pollinators, as well as the amounts of pollen removed and deposited, correlates of male and female fitness (Peter and Johnson 2009). As expected, the success of floral mimics was found to be frequency dependent (Anderson and Johnson 2006). When rewardless mimics were rarer than rewarding plants, they experienced higher rates of pollen removal and deposition than when the mimics were common. When rewardless plants were common, pollinators were less likely to visit a second rewardless plant. It is interesting that populations of rewardless orchids exhibit a high degree of intraspecific variation in the cues that attract visitors (Schiestl 2005). One hypothesis to explain this variation is negative frequency-dependent selection—once a rewardless mimic becomes common, relative to its model, visitors will begin to discriminate against it. Empirical evidence that visitors favor less common morphs of the mimic is equivocal.

Selection analyses have been carried out for a relatively small number of floral traits of a small number of rewardless flowers. For example, taller plants, smaller flowers, and those that opened earlier in the season were favored by selection for a rewardless orchid in Nova Scotia as measured by pollinia removed or deposited (O'Connell and Johnston 1998). Rewardless flowers may take advantage of the innate preferences of flower visitors; orchid flowers that resembled female wasps were much longer and wider and released 10 times as much of the attractive odors as actual female wasps (Schiestl 2004). Male wasps presumably selected for these exaggerated traits.

Rewardless flowers use multiple cues to attract visitors; these mimic the signals of other, more rewarding organisms. For flowers that attract visitors that are foraging for food, visual signals are thought to be most important since insect visitors such as bees search for flowers by associating color with rewards (Galizia et al. 2005, Schiestl 2005). Many rewardless flowers do not mimic specific models but present visitors with cues that are generally attractive (Schluter and Schiestl 2008). In other instances, the cues are attractive to specific visitors as when rewardless flowers emit odors similar to decaying flesh or feces. Flowers attracting male insects by mimicking females often match the specific odors, shape, color, and texture of a receptive model (fig. 7.1) (Schiestl 2005, Schluter and Schiestl 2008).

Rewardless species often match the specific traits of rewarding plants that grow in the same community. Species of South African plants that are pollinated by the long-tongued fly *Prosoeca ganglbaueri* show signs of convergent evolution at a local scale and divergence at a regional scale (Anderson and Johnson 2009). Tongue length of this fly varies from 20 to 50 mm among sites, and this geographic variation is matched by the floral morphologies at these sites. In other words, floral depths of the plant species in the guild converged to match the local tongue length of flies at any given site. The match between tongue length and flower depth affected seed set where it was examined (Anderson and Johnson 2008).

Since visitors to rewardless flowers accrue no benefits, they would be expected to evolve sensory capabilities and behaviors enabling them to avoid rewardless flowers. Why hasn't this selection caused visitors to discriminate against rewardless "cheaters," eliminating them from the population? There are two non–mutually exclusive groups of hypotheses to answer this question.

First, it may be difficult for visitors to distinguish rewarding and unrewarding flowers. Many flowers conceal their rewards, so they become apparent to visitors only after the insect alights on the flower (Bell 1986). Rewardless flowers are unpredictable since flowers will be temporarily empty some of the time on an otherwise rewarding plant, permanently empty for some individual flowers on an otherwise rewarding plant, and permanently empty for all flowers of some plant species. Since flower visitors must constantly deal with this unpredictability, they may be less able to specifically avoid permanently rewardless flowers (Thakar et al. 2003, Willmer 2011). Once the visitors realize that the flower has no rewards, pollen transfer has already occurred in many cases (Johnson 2000).

Second, the benefits of avoiding rewardless flowers may not exceed the costs of visiting them (Schemske et al. 1996). Rewardless flowers (empties) are encountered frequently, but inconsistently, by floral visitors (Smithson and Gigord 2003, Willmer 2011). It may also be more difficult for insects such as bees to learn negative stimuli than positive stimuli (Dukas and Real 1993). For male insects that are attracted to orchid flowers that resemble conspecific females, there may be relatively small costs associated with "copulating" with these flowers. In many cases, males are attracted to these deceitful flowers only before actual females of their own species become available (Willmer 2011). In summary, rewardless flowers are the exception that proves the rule that any floral cue that attracts visitors and is correlated with high plant fitness can be favored by selection.

This chapter has provided some guidelines to evaluate whether traits that currently function in plant sensing and communication can be considered as adaptations. Plants possess traits that are wonderful in their diversity and in the functions that they perform. However, to be considered as adaptations that have been molded by natural selection, those traits must show demonstrable variability that is heritable and be associated with a fitness benefit. This turns out to be a surprisingly tall order.

10 Plant Sensing and Communication in Agriculture and Medicine

10.1 Manipulating the sensing and communication process

All consumers are ultimately dependent on plants for nutrition; without plants little other life can be sustained on the planet. Human omnivores procure a majority of dietary calories and vitamins from plant foods. We also rely on plant products for fiber, fuel, building materials, and medicines. Even as agriculture, medicine, and other human activities change in a technologically dynamic world, our fundamental reliance on plants remains much as it has been for centuries.

Sensing and communication provide plants with information about their external (and internal) environments. Plants process the information contained in cues and signals and allocate resources accordingly. Since cues and signals are decoded by plant tissues, manipulating these cues and signals may provide agriculturalists with an effective means of shaping plants to produce particularly desirable products. It is sometimes argued that natural selection has already produced the optimal solutions to environmental challenges and that further tinkering, "messing with nature," will not be beneficial in the long run. There are at least three problems with this argument. First, natural selection favors those traits that increase the fitness of individual plants, but these may not be optimal from a human point of view. This conflict becomes clear when considering a seedless fruit—it has no value in a traditional selective environment but considerable value to consumers. Second, more generally, natural selection has optimized plants for one suite of environmental conditions. However, modern agriculture often involves rather different conditions so that traits that were selected in nature may not be selected in agriculture. This mismatch becomes particularly relevant in a world where fundamental environmental conditions are changing (Denison et al. 2003). Third, natural selection has been limited by existing variation, but newer techniques allow breeders to produce plants with highly desirable traits that have not been seen before and would not have arisen naturally.

Attempts to use our knowledge of plant sensing and communication to improve agricultural production fall into two general categories. First, genetic modifications alter the abilities of plants to sense environmental cues or their responses. These modifications have traditionally been conducted using heritable variation existing within the gene pool of the crop species. Recently, genetic techniques have allowed us to move genes that code for desirable traits from any species and insert them into crop genomes. These techniques certainly have limitations (e.g., it is easier to insert single traits) but also hold enormous potential. These new genetically modified plants may have highly desirable traits, although they also pose ethical and safety concerns (Gilbert 2013).

Second, rather than attempting to change the way that plants sense and respond, it is often possible to improve their phenotypes by manipulating the environmental cues that they receive. Modifications may involve the chemical milieu of individual cells, the particular cues they are exposed to, or larger-scale conditions experienced by the entire plant. For example, one commonly used technique is to apply chemical elicitors that are perceived by plants and induce phenotypes that we find desirable.

This chapter will consider how we can use plant cues and signals to manipulate plant traits to make plants more effective at capturing resources, better protected against herbivores and pathogens, and generally more valuable.

10.2 Manipulating resource acquisition and allocation

Many sun-adapted plants respond to light quality to adjust their morphologies, the shade avoidance responses that are well characterized (see section 5.2.1). These responses help individual plants avoid shade but may not be conducive to high yields in dense agricultural plantings (Smith 1992, Denison et al. 2003, Denison 2012). Shade avoidance responses shunt resources to vertical growth of stems at the expense of leaves, roots, and reproductive structures. Thus, shade avoidance responses have been associated with reductions in crop yields in many instances (Smith 1992, Morgan et al. 2002). Stands of homogeneous crop plants experience all of the costs and none of the benefits of attempting to overtop their neighbors. Most crops are shade avoiders, in contrast to shade-tolerant species that exhibit fewer and weaker shade avoidance responses and are more efficient at growing under low-light conditions (Smith 1992, Kebrom and Brutnell 2007, Gommers et al. 2013).

Intense breeding during domestication reduced the strength of shade avoidance responses in many crop species (Smith 1992, Kebrom and Brutnell 2007). Many of the grain varieties associated with the "green revolution" exhibit reduced stem growth and invest more in reproduction (Morgan et al. 2002, Denison et al. 2003). In addition to conventional breeding efforts, overexpression of phytochrome light receptors has been found to suppress the shade avoidance responses in several crop plants including tomato, tobacco, potato, and rice. Genetically engineered rice with suppressed shade avoidance responses produced more panicles and greater grain yields (Garg et al. 2006). Other genetic modifications that alter the perception of light or plant responses could improve crop yields in the future (Smith 1992). These modifications have the potential to do what natural selection cannot: favor reproduction of the monoculture or community of agricultural plants over the success of individuals (Denison et al. 2003, Denison 2012).

10.3 Manipulating tolerance to abiotic stress

Phytochrome light receptors influence a wide variety of plant behaviors in addition to shade avoidance responses. Artificial manipulation of the light receptor systems may allow control of many of these behaviors. For example, aspen trees that overexpressed genes coding for phytochrome A from oats experienced altered detection of photoperiod and improved seasonal acclimation to cold stress (Olsen et al. 1997). Acclimation to cold stress in several plants, including *Arabidopsis*, involves induction of transcription factors (CBFs) (Thomashow 1999). Similar responses have been reported for other plant species that acquired tolerance to drought and salinity stress (Chinnusamy et al. 2010). Genetic manipulation of the CBF pathway can improve tolerance to cold and other stresses and many crops have now been engineered for greater cold tolerance (Chinnusamy et al. 2010, Sanghera et al. 2011). However, these approaches to stress tolerance have not yet been incorporated into production agriculture.

Plant responses to many stresses depend upon their previous experiences (see chapter 3). We can take advantage of this phenomenon by priming plants with chemical elicitors, making them respond more rapidly and effectively when they encounter the actual stressful conditions (Conrath et al. 2006). For example, pretreating *Arabidopsis* plants with the nonprotein amino acid BABA made them more tolerant of drought and salt stress (Jakab et al. 2005). When pretreated plants were faced with these stresses, they responded by producing more abscisic acid, which in turn led to greater expression of genes associated with stress tolerance and more rapid stomatal closure. Evidence also suggests that pretreating with jasmonates alleviates salt stress in several crop species (Fedina and Tsonev 1997, Shahbaz et al. 2012). Several commercial groups are developing products termed "plant health regulators" that can be applied to prime plants for tolerance to abiotic stresses. For example, tomato plants that were pretreated with the activator "Alethea," which includes jasmonates, salicylates, and arginine, became more tolerant of salinity stress (Wargent et al. 2013). These techniques are being explored but have not yet been used in production agriculture.

10.4 Manipulating resistance to pathogens

Plants naturally respond to cues associated with pathogen attack by upregulating genes that will make them more resistant or more tolerant or by priming that will allow them to respond to actual pathogen attack (see section 8.2). The earliest attempts to use these techniques involved vaccinating plants with avirulent microbial strains to gain protection against economically damaging pathogens (Chester 1933, Kuc 1987). These vaccination techniques were effective in some situations although they were not widely adopted because they were not commercially profitable (Kuc 1995). Even when they were cost-effective for growers, they were not easily patented and marketed, so agribusiness was not interested in providing them.

There have been several commercial attempts to introduce rhizobacteria to crop plants in order to induce systemic resistance against a variety of viral, bacterial, and fungal diseases in production agriculture (reviewed in Vallad and Goodman 2004). Agricultural uses have been associated with reduced disease symptoms and increased crop yields. Beneficial rhizobacteria stimulate pathways regulated by jasmonate and ethylene (Knoester et al. 1998). In addition, several products have been marketed that contain harpin, an elicitor found in bacteria that plants recognize. Other, less well-defined, extracts of bacteria are reported to induce resistance and have been sold commercially in Europe (von Rad et al. 2005).

Fungal cell walls, as well as exoskeletons of crustaceans, contain linear polysaccharides known as chitosan (Hadwiger 2013). These chemicals are extracted from waste products of crab and shrimp exoskeletons and have been used as a seed treatment. Chitosan is recognized by many crop plants, causing them to induce resistance against fungal diseases such as straw breaker fungus in wheat. Chitosan is less expensive to use than competing chemical pesticides although it provides less complete protection and can cause other complications associated with harvesting.

An alternative to introducing biological agents involves identifying the cues with which plants recognize pathogens or the signals that plants use to coordinate their responses; these chemicals can then be used to manipulate levels of plant resistance. Many different chemicals have now been identified that elicit nonspecific induction of resistance against pathogens (Lyon 2007). Salicylic acid (SA) and its analog, 2,6-dichloroisonicotinic acid (INA), are potent inducers of resistance against most of the pathogens that are affected by the SA pathway (Kessmann et al. 1994). These products are effective inducers of resistance but also cause many undesirable side effects that are also mediated by the SA pathway, making them impractical for widespread use (Ryals et al. 1996). In some instances they caused phytotoxicity and in others they provided control of pathogens but reduced yields.

The most widely used artificial elicitor is benzo (1,2,3) thiadiazole-7-carbothioic acid S-methyl ester (BTH), which has activity against a range of diseases in many different crops (Vallad and Goodman 2004). It has been marketed by Syngenta as Actigard in the United States and as Bion in Europe. BTH is an analog of salicylic acid and induces systemic acquired resistance, reducing disease symptoms and increasing yields. It is particularly effective against mildew in wheat but has also been found to provide resistance against numerous diseases and other pests ranging from whiteflies to parasitic plants. However, the commercial success of BTH has been limited. BTH must be applied before infestation and has been disappointing to growers who have become accustomed to the dramatic and rapid effects of commercial pesticides. Several other elicitors of resistance against various pathogens have been marketed commercially although their use has been much more limited than BTH (Lyon 2007).

A potential drawback of using elicitors to artificially induce resistance against pathogens is that resistance is likely to be costly to the plants, particularly when the risk of attack is low. *Arabidopsis* plants that were treated with exogenously supplied SA or BTH had reduced growth and seed production when pathogens were not present (Cipollini 2002, van Hulten et al. 2006). In contrast, priming plants with relatively low doses of elicitors resulted in no detectable decrease in growth or seed set in the absence of pathogens (van Hulten et al. 2006). Priming allowed plants to save the costs when a response was not needed but also allowed them to respond more rapidly and more effectively to pathogen attack (Conrath et al. 2006). *Arabidopsis* plants that were primed by elicitors experienced approximately 50% more seed production than controls when pathogenic bacteria were present (van Hulten et al. 2006).

By identifying and mimicking the signals and cues used by plants, agriculturalists can manipulate natural plant defenses in accordance with the perceived risks. Priming plants with relatively low doses of elicitors is likely to be more effective than inducing direct defenses with higher doses. Applying these techniques effectively to growing plants in production agriculture may require considerable knowledge of the specific crop and situation. Treating seeds with low doses of elicitors may prime them without requiring early detailed knowledge of pathogen risk. Tomato seeds treated with elicitors had more effective responses to pathogens for 8 weeks without experiencing reductions in growth when pathogens were not present (Worrall et al. 2012). As discussed above, elicitors that prime plant defenses are unlikely to provide the same dramatic results that chemical biocides have provided in the past.

Since defenses are likely to be costly for plants to employ, growers will want to induce them only when the risk of attack by an economically damaging pathogen is high. As with any other tool in integrated pest management, induced host plant resistance will be more effective if growers can predict the likelihood of future risk. This process will be facilitated by chemical markers that reliably indicate the presence of particular pathogens. Biosensors based on insect antennae have been developed that show great sensitivity and specificity to volatiles released by specific pathogens (Schutz et al. 1999). More recent efforts allow the volatiles to be collected in the field and analyzed by gas chromatography (Jansen et al. 2009, Laothawornkitkul et al. 2010). Considerably more work is needed before these sensors can become useful tools for agriculture.

Production in many natural and agricultural plant communities is limited by the availability of nitrogen. The green revolution has increased crop yields throughout the world by augmenting natural levels of nitrogen with synthetic fertilizers. This practice is expensive and unsustainable, and it causes considerable environmental harm. Natural selection has solved this problem for some plants by allowing them to take advantage of bacteria that fix nitrogen from the atmosphere. Plants are normally well defended against bacterial infections, but some plants engage in a complex dialogue with certain bacteria, allowing those strains to pass through the plant's immunological barriers and to become established within plant tissues (see section 8.5). Unfortunately, the grasses that are most important as sources of food do not naturally form these associations. In has long been recognized that overcoming this limitation would be an enormous boon to humankind (Burrill and Hansen 1917).

Legumes and a few other plants form mutualistic associations with bacteria that begin with mutual recognition and generally involve the plants producing nodules. The bacteria produce a nodulation factor (Nod) that is recognized by the plants and causes them to form nodules which become the homes for the invading bacteria (Beatty and Good 2011). Recently, it has been discovered that nonlegumes also recognize the Nod factor of bacteria and suppress the plant's immune responses, albeit not as effectively as with legumes (Liang et al. 2013). In addition, the legume receptors that suppress immune responses have been identified. These developments lend support to the hope that cereals can be engineered to house N-fixing bacteria. However, this will be a tall order because the symbiosis completely restructures the plant cell, creating an environment with low oxygen, and many factors must be present for this to work.

10.5   Manipulating resistance to herbivores

Attempts to control herbivores by manipulating host plant resistance have been conceptually similar to those used against plant pathogens. Entomologists became aware of induced resistance decades after their colleagues in plant pathology. Green and Ryan (1972) found that when Colorado potato beetles wounded the leaves of tomato plants, concentrations of proteinase inhibitors increased; these compounds were presumed to defend the plants against subsequent herbivory. Haukioja and Niemela (1977) found that caterpillars reared on leaves from birch trees that had been previously defoliated grew more slowly than those from undefoliated trees. This response was hypothesized to provide a potential explanation for cyclic outbreaks of forest insects (Haukioja and Hakala 1975).

Early attempts to use induced resistance to control agricultural pests relied on introducing herbivores that were not economically damaging to protect crops against those that were. For example, vineyards that had experienced chronic infestations of economically damaging Pacific spider mites could be protected against these invasive pests by reintroducing or favoring native Willamette mites, another herbivore (Karban, English-Loeb, and Hougen-Eitzman 1997). Willamette mites made vines less favorable hosts for Pacific mites and a single "vaccination" reduced pest populations and increased yields and grape quality (English-Loeb et al. 1993). The two mite species are difficult to distinguish in the field, and to provide effective control, Willamette mites must be present early in the season as new shoots are expanding (Hougen-Eitzman and Karban 1995). Vaccinations never became widely used because there were no easy opportunities for agricultural companies to market the technique.

Many of these practical difficulties may be alleviated by applying elicitors of induced resistance or by priming plants against herbivores. Tomato plants that were sprayed in the field with low concentrations of jasmonic acid or methyl jasmonate induced production of several chemicals associated with defense against herbivores (Thaler et al. 1996). Plants that were treated with these signals were less suitable hosts for foliage-feeding herbivores and received 60% less damage to leaves than untreated controls (Thaler 1999b). Neither costs of inducing resistance nor benefits of reduced damage in terms of yield were not found in this study.

In addition to direct effects of induced resistance, some plants may gain protection by attracting the predators and parasitoids of their herbivores (see section 6.3). Damaged plants emit relatively high concentrations of volatiles that may provide useful information to searching carnivores (Vet and Dicke 1992). Emissions that attract enemies of herbivores are controlled by cues and signals and may be manipulated by artificial applications of elicitors. For example, caterpillars in agricultural fields suffered twice the rate of parasitism by wasps near tomato plants that had been exposed to artificially supplied jasmonate cues as did controls (Thaler 1999a).

Workers have attempted to increase the numbers and the effectiveness of predators and parasites by attracting them to a variety of crops using artificial emitters of plant cues or their analogues (Kaplan 2012). At least two synthetic products are commercially marketed to attract predators (Preda-

Lure and Benallure); these rely on methyl salicylate and 2-phenylethanol. Controlled release dispensers can attract predators into crop fields, reestablishing ratios of predators and prey that have become depleted by other agricultural practices. For example, dispensers that released methyl salicylate attracted predatory insects in hop and grape vineyards and these enhanced densities of predators were associated with reduced numbers of pest spider mites at the scale of entire fields (James and Price 2004). Fewer studies have considered whether predators will remain in fields, and results have been mixed (Kaplan 2012). It is not known whether attracting carnivores can be an effective strategy over larger spatial scales or whether carnivores will become habituated and less responsive to high concentrations of attractive compounds over longer time frames. Predators and parasites use reliable plant cues to locate their hosts. Manipulating plants to emit these cues when hosts are not actually abundant reduces the information value of the cues and may select for predators and parasites that ignore those cues. Some herbivores also use these volatiles to locate suitable hosts so that artificially increasing the emission rates can result in increased levels of plant damage (von Merey et al. 2011). While several studies have found reduced pest populations associated with artificial releases of attractive cues, beneficial effects on plant yields have not been demonstrated.

Some plants constitutively produce compounds that are attractive to the predators and parasites of herbivores (Khan et al. 1997). For example, molasses grass and *Desmodium* legumes in east Africa emit volatiles that repel stem-boring herbivores and attract parasitoid wasps. These plants also emit other chemicals that are typically released by maize after it has been damaged by chewing herbivores (Khan et al. 2008). Intercropping fields of maize or sorghum with molasses grass or *Desmodium* has been found to reduce herbivore numbers and increase carnivores. Trap plants that release green leaf volatiles and are more attractive to the herbivores than the crop species may also be included in the intercrop mix. Stem-boring herbivores failed to complete development in these attractive but unsuitable hosts (Khan et al. 2008). By manipulating the cues that attract and repel herbivores and carnivores, it has been possible to employ "push-pull" strategies to effectively control pest problems; this method has been widely adopted by many small growers in Africa (Cook et al. 2007, Khan et al. 2010).

These strategies rely on using chemical elicitors to change the cues emitted by crop plants or on changing the chemical environment that surrounds the crop. Another possible strategy involves genetically manipulating the cues that the crop emits. Since many herbivores locate their hosts using host-specific cues, it may be possible to alter these cues and reduce the attractiveness of crops. This can be accomplished with traditional plant breeding or more rapidly with genetic engineering (Bruce 2010). New techniques allow desirable traits from any organism to be inserted into the crop genome and expressed. Plants have been genetically modified to produce the alarm pheromone of aphids (Beale et al. 2006). This chemical causes aphids to stop feeding, disperse, and frequently to drop from their host plant. Mint plants naturally produce the compound, and modified *Arabidopsis* and wheat that express it have been created (Pickett et al. 2014). Parasitoids of aphids were also more attracted to plants that produced the alarm pheromone, although emissions will probably need to be made inducible before it has a real chance of widespread use.

Other plants have been altered to become more attractive to the predators and parasites of pests. Genetically engineered *Arabidopsis* with a synthase from strawberry had altered expression of terpenes and recruited more predatory mites (Kappers et al. 2005). Similarly, *Arabidopsis* plants that were engineered to overexpress terpene synthase or green leaf volatiles became more attractive to parasitic wasps (Schnee et al. 2006, Shiojiri et al. 2006). Rice plants that overexpressed (E)-β-caryophyllene synthase became more attractive to parasitoids of the rice brown planthopper, a major pest. Maize plants that were transformed to express (E)-β-caryophyllene were more attractive to entomopathogenic nematodes, which reduced local infestations of corn rootworms (Degenhardt et al. 2009). The ability to express this volatile was present in the ancestral maize lines but was subsequently lost by breeding efforts; it could be restored by either classical breeding methods or by genetic engineering.

Many researchers have now altered the expression of volatile emissions, using both overexpression and knockouts, and found changes in responses of herbivores and carnivores (Kant et al. 2009). These genetic alterations have made it clear that even simple manipulations have far-reaching consequences and often unintended phenotypic results (Dudareva and Pichersky 2008). For example, overexpression of genes producing linalool resulted in a phenotype with reduced growth that was transmitted to subsequent generations (Aharoni et al. 2003). These unwanted consequences can be avoided in most cases and should not detract from the potential of genetic modifications. However, these techniques have not yet been used successfully in commercial agriculture. Employing them without carefully considering the selection pressures that they impose on all interactors can be misguided.

Much of plant reproduction is orchestrated in response to cues and signals. First, the timing of flowering and fruiting is controlled by environmental cues such as photoperiod and temperature. Plants assess photoperiod largely by their perception of the length of the night. Next, flowers advertise rewards and attract visitors that, in some cases, pollinate them. Similarly, fruits use cues and signals to attract consumers that, in some cases, disperse seeds to locations favorable for growth. Finally, germination occurs when seeds respond to cues correlated with environmental conditions that will be favorable for growth. As consumers of plants, we can improve the quality, timing, and quantity of "food" produced by plants by controlling these various steps in the communication process.

Controlling the timing of reproduction can allow growers to produce fruits out of season, when demand is particularly high. Buds of deciduous trees normally go through a rest period in winter that can be shortened by subjecting them to various combinations of conditions, including quicker photoperiodic cycles, chilling, desiccation, and defoliation (Erez et al. 1966, Erez 1987). These environmental cues can be manipulated in orchards in tropical environments or for containerized plants in greenhouses to produce two crops of peaches per year (Sherman and Lyrene 1984, George et al. 1988). Peach trees were introduced to Venezuela by Spanish colonists; after approximately ten generations of selection for early flowering and fruit ripening, coupled with two dry and two wet seasons per year, the trees began producing crops of peaches semiannually. Feral peach trees near Brisbane, Australia, also produce two crops in many years when drought interrupts the growing seasons. Containerized peach trees in Israel have been induced to produce two crops by accelerating seasonal cycles of photoperiod and temperature.

These techniques involve altering the environmental cues that plants perceive, either naturally or artificially. Most recent efforts have concentrated on genetically manipulating plants to sense and respond differently to environmental cues. For example, genes have been identified in sorghum that delay flowering until very late in the growing season, and flowering in wheat and barley can be either delayed or accelerated (Morgan et al. 2002). Traditional crop breeding has already selected for plants with novel reproductive innovations such as greatly accelerated flowering, either reduced lateral branching of flowering stalks (e.g., maize) or increased lateral branching (e.g., rice), and many others. The cues "advertising" modern crop varieties

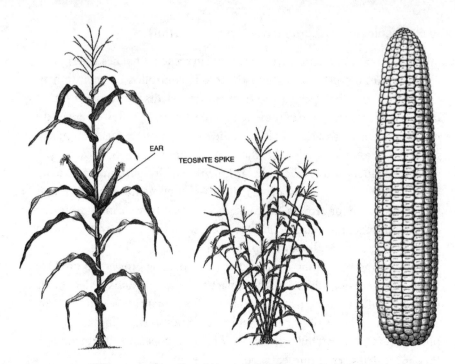

FIGURE 10.1 Differences between the wild teosinte plant and modern domesticated corn.
A. Modern corn is a taller plant with a single unbranched stem that bears a small number of
very large ears. Teosinte is shorter, with many branches that each bear tiny spikes at each node.
B. A spike of teosinte bears a single row of small hard kernels. An ear of modern corn bears
many rows of large soft kernels. Both ears are to scale and are shown at one third their actual
size. Figures from Beadle (1980).

make them almost unrecognizable as descendents of early accessions. For
example, the edible parts of early progenitors of maize were too small to be
eaten directly (Doebley 2004). A spike of teosinte is several orders of magnitude smaller than an ear of domesticated corn, and the morphologies of the
plants are markedly different as well (fig. 10.1, Beadle 1980). Identification of
genes associated with other desirable traits will allow many more improve-
ments in crop species.

Tomato has become the model system for genetically modifying cues
that fruit consumers are likely to find attractive. In the 1980s scientists took
advantage of the unusual way in which *Agrobacterium tumefaciens* inserts a
piece of bacterial DNA when it infects its hosts (Gelvin 2003). Using *Agrobacterium* allowed scientists to introduce foreign DNA into the tomato genome, where it was then expressed. The first commercially available food

that had been genetically modified was a tomato that exhibited delayed ripening (Martineau 2001). Mature tomatoes are considered ripe after 45–55 days, and these fruits undergo a suite of changes that are controlled by ethylene signals including a change of color from green to red, softening of the fruit, and modifications in taste and odor. The ripening process can be controlled by transforming any of several enzymes that are involved in the pathway that produces ethylene or by reducing the ability of the fruit to perceive or respond to the ethylene cues. It has also been possible to control the breakdown of pectin in the cell wall of the fruit, essentially slowing down the rate at which the ripe fruit becomes soft, facilitating fruit handling and transport (Kramer et al. 1992). The Flavr Savr tomato, developed by Calgene, relied on this latter modification. Unfortunately, transformed Flavr Savr tomatoes possessed neither improved taste nor aroma (R. Karban, personal observation). Tomatoes with delayed ripening were marketed in the United States beginning in 1994 and in Europe in 1996. Business difficulties, independent of the product, and pressure from consumer advocacy groups caused the companies that supplied transgenic tomato seeds to discontinue production in the U. S. in 1997 and in Europe in 1999 (Martineau 2001). Tomato plants have been used in experiments that have attempted to add other desirable traits. Genes for cold tolerance from flounders and genes for salt and drought tolerance from rice, *Arabidopsis*, and tobacco have been inserted into the tomato genome, although these plants have not reached markets (Zhang and Blumwald 2001, Vannini et al. 2007, Lemaux 2008, Goel et al. 2010). Genes for resistance against insects and pathogens have also been inserted into tomato, although these kill the pests rather than altering the signaling process and will not be described further here.

The volatiles emitted by tomato fruits have been modified on several occasions by introducing genes from other species (Speirs et al. 1998, Lewinsohn et al. 2001). These changes were reported to give the tomatoes a more intense ripe fruit flavor. A gene from lemon basil has been inserted into tomato, improving its taste and smell to a majority of people tested, but reducing its nutritional value (Davidovich-Rikanati et al. 2007). Other researchers have inserted the genes associated with increased production of anthocyanins, which are thought to have health benefits as antioxidants that may prevent cancer (Zuluaga et al. 2008).

Researchers have also begun transforming other plants to produce advertisements that we find attractive or useful. For example, transgenic peppermint plants produce as much as 78% more essential oils than the natural hybrid plants that are used in commercial agriculture; the relative blend of essential oils can also be controlled by these manipulations (Lange et al. 2011).

10.7   As a source of medicines

The discovery, production, and use of antibiotics are probably the main accomplishments of Western medicine. Of course, antibiotics have existed and been used by microbes for most of the history of life, but the discovery of penicillin marked a turning point in the history of our species. At various times in the past, humans have inferred that molds were useful in combating infections; Imhokep, an Egyptian healer, applied moldy bread to treat surface wounds, and various other ancient traditions called for the application of moldy grains for therapeutic purposes (Wainwright 1989). Alexander Fleming (1929) postulated that antibacterial effects associated with fungi were chemically mediated and that these compounds could be co-opted for chemotherapy. While most competitive interactions between microbes cannot be considered communication because the responder does not choose whether to respond (see definition in chapter 1), nonetheless interactions between plants and other organisms can provide sources of new drugs and therapies. Indeed, a majority of new medicines approved for use between 1980 and 2010 were derived from natural products, particularly compounds that were originally involved in interactions among microbes and/or plants (Newman and Cragg 2012).

One example of pharmaceutical exploitation of possible signaling between plants and herbivores involves the phytochemical artemisinin. Artemisinin is a sesquiterpene lactone produced by *Artemisia annua* that protects plants against attack by herbivores (Ferreira and Janick 1995, Maes et al. 2011). It also has activity against diverse plant pathogens including viruses, bacteria, and fungi such as *Verticillium dahliae* (Tang et al. 2000, Efferth 2009). Artemisinin has been used in Chinese medicine since at least 168 B.C. and is currently used as the drug of choice against malaria, particularly malarial strains that are resistant to other antibiotics (Efferth 2009). Artemisinin has also been shown to have activity against some forms of cancer, parasitic worms such as those that cause schistosomiasis, and viral infections such as hepatitis B and C. Artemisinin can be autotoxic to plants but is safely sequestered in glandular trichomes; inflorescences contain 10 times the concentration of leaves on a dry weight basis, and concentrations in trichomes peak during anthesis (Ferreira and Janick 1995). Consistent with the hypothesis that artemisinin evolved as a defensive chemical, the jasmonate pathway was found to regulate its synthesis (Maes et al. 2011). Experimental elicitation with jasmonate cues increased the density of trichomes, their size, and their production of artemisinin.

Another interesting example involves the grass *Vetiveria zizanioides*, which produces essential oils that are highly valued in cosmetics and also show antimicrobial, insecticidal, and antioxidant activity (Del Guidice et al. 2008). The oils are made up of a complex blend of sesquiterpenes and alcohols and are produced in specialized secretory cells surrounding the roots. The oils are released in response to insect feeding, application of insect regurgitant, or wounding, and they are repellent to root-feeding nematodes and many other microbes. Bacteria were found in the oil-producing root cells, and plants grown without these bacteria produced far less oil. These bacteria can grow using vetiver terpenes as a food source, and they induce gene expression of sesquiterpene synthase in the plant roots. While the detailed biology of this relationship still needs to be elucidated, it is clear that this plant communicates with its associated root bacteria and that their interaction is involved in the production of chemicals that are valuable to humans.

Another example involves the tree resins that have been combined into the medicinal product Tiger Balm. Many of the trees in Southeast Asian tropical forests produce aromatic resins and oils that apparently target animal thermosensors (Maffei et al. 2011). Camphor and menthol affect transient receptor potential channels that transmit pain signals to the brains of mammals (Xu et al. 2005, Macpherson et al. 2006, Vogt-Eisele et al. 2007). Rubbing them into a locally painful area may disrupt this signal, provide the sensation of cooling, and mildly irritate the skin, thus increasing blood flow.

The list of examples of plant cues that are sensed by other organisms and also have valuable medicinal properties is extensive (Maffei et al. 2011). For example, most essential oils of the varied plant products that we consider spices are antibacterial and help protect against symptoms caused by *Salmonella* (Kobilinsky et al. 2007).

10.8 Plant sensing in the future of humankind

As the human population continues to increase, we will require a larger supply of food and other resources. Providing for these needs will take up much of the space on our planet, and we can expect rates of extinctions of other species including plants to continue to accelerate (van Vuuren et al. 2006, Pereira et al. 2010, Dawson et al. 2011). As a consequence we will lose the genetic plant diversity that contains new foods, medicines, and other useful products.

Our actions also have more immediate impacts on how plants will sense their future environments and communicate. For example, there is growing evidence that atmospheric ozone, a common pollutant, may interfere with plant volatile communication. Plants that are attacked by herbivores emit green leaf volatiles and other terpenoids that attract predators and parasitoids of the herbivores (see section 6.3). These attractive volatiles were rapidly oxidized by high, but realistic, levels of ozone although wasps were apparently able to use other, more stable cues to locate herbivore hosts (Pinto et al. 2007). Elevated CO_2 also influenced volatile emissions of damaged plants and may affect communication between plants and carnivorous insects (Vuorinen et al. 2004). Both a generalist predator and a specialist parasitoid were attracted to damaged cabbage plants only at ambient CO_2 levels and not at elevated levels.

Ozone may also affect communication between plant tissues. High ozone concentrations (80 ppb) commonly encountered today reduced the distance over which volatile signaling occurred among lima bean plants (Blande et al. 2010). Emitter plants were infested with spider mites and plant responses were assayed by the production of extrafloral nectar in receiver plants. Ozone degrades many of the volatiles emitted by plants attacked by herbivores although the precise identity of the volatile cues remains unknown (Blande et al. 2011, Holopainen and Blande 2012). Higher ozone concentrations caused plants to constitutively produce extrafloral nectar, a response usually associated with attack by herbivores (Blande et al. 2010).

Plants have long been a source of inspiration for humankind from the development of agriculture to space travel (Benyus 1997). We have mimicked plant adaptations and used them to develop new technologies. Mimicking plants successfully requires a deep understanding of how they work.

For example, the Swiss engineer George de Mestral patented Velcro in 1955 after repeatedly removing the seed pods of burdock (*Arctium* sp.) from the fur of his dog. Close inspections revealed that the seed pods were covered by hundreds of small hooks that attached to any loops such as those found on cloth or dog fur. Hooks on the pods made communication unnecessary for successful dispersal by animal vectors. This observation inspired a method to reversibly provide a strong attachment between two items, which de Mestral was able to recreate with synthetic materials (Strauss 2002). Velcro has been used for many purposes including holding the heart together during the first successful artificial heart surgery and by NASA to secure astronauts' space suits and to keep objects from floating away in near-weightless conditions.

Beyond useful gadgets, plants may provide clues about harnessing the sun's energy in solar cells (Benyus 1997). Photosynthetically active pigments and man-made solar cells both capture light energy and store it. However, plants and especially photosynthetic bacteria are many times more efficient at this process than are the best devices that we have thus far managed to create. Plants have been exposed to millions of years of natural selection to solve some of the same problems that modern humans are facing for the first time (Denison 2012)

This book began with the observation that plants and animals had very different sensory abilities and potentials for communication. Plants lack central nervous systems and are probably unable to reason in a way that resembles the process that animals exhibit. And yet, they perceive many of the same environmental cues as animals do, sometimes with greater acuity. In some cases plants perceive cues that humans cannot. They store and process information, focus attention, and respond in highly ordered ways that increase their Darwinian fitness. Far from being inanimate, plants sense, communicate, and show sophisticated conditional behaviors that we are just starting to appreciate.

References

Abeles, F. B., P. W. Morgan, and M. E. Saltveit. 1992. Ethylene in Plant Biology, 2nd ed. Academic Press, San Diego.

Ackerman, J. D. 1986. Mechanisms and evolution of food-deceptive pollination systems in orchids. Lindleyana 1:108–113.

Adler, F. R., and D. Grunbaum. 1999. Evolution of forager responses to inducible defesnes. Pages 259–285 *in* R. Tollrain and C. D. Harvell, editors. The Ecology and Evolution of Inducible Defenses. Princeton University Press, Princeton, NJ.

Adler, F. R., and R. Karban. 1994. Defended fortresses or moving targets? Another model of inducible defenses inspired by military metaphors. American Naturalist **144**:813–832.

Adler, L. S. 2000. The ecological significance of toxic nectar. Oikos **91**:409–420.

Adler, L. S., and R. E. Irwin. 2005. Ecological costs and benefits of defenses in nectar. Ecology **86**:2968–2978.

Adler, L. S., M. G. Seifert, M. Wink, and G. E. Morse. 2012. Reliance on pollinators predicts defensive chemistry across tobacco species. Ecology Letters **15**:1140–1148.

Agrawal, A. A. 1998a. Induced responses to herbivory and increased plant performance. Science **279**:1201–1202.

Agrawal, A. A. 1998b. Leaf damage and associated cues induce aggressive ant recruitment in a neotropical ant-plant. Ecology **79**:2100–2112.

Agrawal, A. A. 2005. Future directions in the study of induced plant responses to herbivory. Entomologia Experimentalis et Applicata **115**:97–105.

Agrawal, A. A., and R. G. Colfer. 2000. Consequences of thrips-infested plants for attraction of conspecifics and parasitoids. Ecological Entomology **25**:493–496.

Agrawal, A. A., and B. J. Dubin-Thaler. 1999. Induced responses to herbivory in the Neotropical ant-plant associations between *Azteca* ants and *Cecropia* trees: responses of ants to potential inducing cues. Behavioral Ecology and Sociobiology **45**:47–54.

Agrawal, A. A., and R. Karban. 1997. Domatia mediate plant-arthropod mutualism. Nature **387**:562–563.

Agrawal, A. A., E. E. Kearney, A. P. Hastings, and T. E. Ramsey. 2012. Attenuation of the jasmonate burst, plant defensive traits, and resistance to specialist monarch caterpillars on shaded common milkweed (*Asclepias syriaca*). Journal of Chemical Ecology **38**:893–901.

Agrawal, A. A., C. Laforsch, and R. Tollrian. 1999. Transgenerational induction of defences in animals and plants. Nature **401**:60–63.

Agrawal, A. A., and M. T. Rutter. 1998. Dynamic anti-herbivore defense in ant-plants: the role of induced responses. Oikos **83**:227–236.

Aharoni, A., A. P. Giri, S. Deuerlein, W.-J. de Kogel, F. W. A. Verstappen, H. A. Verhoeven, M. A. Jongsma, W. Schwab, and H. J. Bouwmeester. 2003. Terpenoid metabolism in wild-type and transgenic Arabidopsis plants. Plant Cell **15**:2866–2884.

Aizenberg-Gershtein, Y., I. Izhaki, and M. Halpern. 2013. Do honeybees shape the bacterial community composition in floral nectar? PLoS One 8:e67556.

Akiyama, K., K. Matsuzaki, and H. Hayashi. 2005. Plant sesquiterpenes induce hyphal branching in arbuscular mycorrhizal fungi. Nature **435**:824–827.

Alborn, H. T., T. C. J. Turlings, T. H. Jones, G. Stenhagen, J. H. Loughrin, and J. H. Tumlinson. 1997. An elicitor of plant volatiles from beet armyworm oral secretion. Science **276**:945–949.

Ali, J. G., and A. A. Agrawal. 2012. Specialist versus generalist insect herbivores and plant defense. Trends in Plant Science **17**:293–302.

Ali, M., K. Sugimoto, A. Ramadan, and G. Arimura. 2013. Memory of plant communications for priming anti-herbivore responses. Scientific Reports **3**:1872.

Allison, J. D., and J. D. Hare. 2009. Learned and naive natural enemy responses and the interpretation of volatile organic compounds as cues or signals. New Phytologist **184**:768–782.

Amasino, R. M. 2004. Vernalization, competence, and the epigenetic memory of winter. Plant Cell **16**:2553–2559.

Anderson, B., and S. D. Johnson. 2006. The effects of floral mimics and models on each others' fitness. Proceedings of the Royal Society B **273**:969–974.

Anderson, B., and S. D. Johnson. 2008. The geographical mosaic of coevolution in a plant-pollinator mutualism. Evolution **62**:220-225.

Anderson, B., and S. D. Johnson. 2009. Geographical covariation and local convergence of flower depth in a guild of fly-pollinated plants. New Phytologist **182**:533-540.

Anderson, S. S., K. D. McCrea, W. G. Abrahamson, and L. M. Hartzel. 1989. Host genotype choice by the ball gallmaker *Eurosta solidaginis* (Diptera: Tephritidae). Ecology **70**:1048–1054.

Appel, H. M., and R. B. Cocroft. 2014. Plants respond to leaf vibrations caused by herbivore chewing. Oecologia **175**:1257-1266.

Arikawa, K. 2003. Spectral organization of the eye of a butterfly, Papilio. Journal of Comparative Physiology A **189**:791–800.

Arimura, G., C. Kost, and W. Boland. 2005. Herbivore-induced, indirect plant defences. Biochemica et Biophysica Acta **1734**:91–111.

Arimura, G., R. Ozawa, J. Horiuchi, T. Nishioka, and J. Takabayashi. 2001. Plant-plant interactions mediated by volatiles emitted from plants infested by spider mites. Biochemical Systematics and Ecology **29**:1049–1061.

Arimura, G., R. Ozawa, T. Nishioka, W. Boland, T. Koch, F. Kuhnemann, and J. Takabayashi. 2002. Herbivore-induced volatiles induce the emission of ethylene in neighboring lima bean plants. Plant Journal **29**:87–98.

Arimura, G., K. Tashiro, S. Kuhara, T. Nishioka, R. Ozawa, and J. Takabayashi. 2000. Gene responses in bean leaves induced by herbivory and by herbivore-induced volatiles. Biochemical and Biophysical Research Communications **277**:305–310.

Armbruster, W. S. 1997. Exaptations link evolution of plant-herbivore and plant-pollinator interactions: a phylogenetic inquiry. Ecology **78**:1661–1672.

Armbruster, W. S. 2002. Can indirect selection and genetic context contribute to trait diversification? A transition-probability study of blossom-colour evolution in two genera. Journal of Evolutionary Biology **15**:468–486.

Armbruster, W. S., J. J. Howard, T. P. Clausen, E. M. Debevec, J. C. Loquvam, M. Matsuki, B. Cerndolo, and F. Andel. 1997. Do biochemical exaptations link evolution of plant defense and pollination systems? Historical hypotheses and experimental tests with Dalechampia vines. American Naturalist **149**:461–484.

Armbruster, W. S., J. Lee, and B. G. Baldwin. 2009. Macroevolutionary patterns of defense and pollination in Dalechampia vines: adaptation, exaptation, and evolutionary novelty. Proceedings of the National Academy of Sciences **106**:18085–18090.

Artus, N. N., M. Uemura, P. L. Steponkus, S. J. Gilmour, C. Lin, and M. F. Thomashow. 1996. Constitutive expression of the cold-regulated *Arabidopsis thaliana* COR15a gene affects both chloroplast and proplast freezing tolerance. Proceedings of the National Academy of Sciences **93**:13404-13409.

Asai, T., G. Tena, J. Plotnikova, M. R. Willmann, W.-L. Chiu, L. Gomez-Gomez, T. Boller, F. M. Ausubel, and J. Sheen. 2002. MAP kinase signalling cascade in Arabidopsis innate immunity. Nature **415**:977–983.

Aschan, G., and H. Pfanz. 2003. Non-foliar photosynthesis: a strategy of additional carbon acquisition. Flora **198**:81–97.

Assmann, S. 1993. Signal transduction in guard cells. Annual Review of Cell Biology **9**:345–375.

Avdiushko, S. A., K. P. C. Croft, G. C. Brown, D. M. Jackson, T. R. Hamilton-Kemp, and D. F. Hildebrand. 1995. Effect of volatile methyl jasmonate on the oxylipin pathway in tobacco, cucumber, and Arabidopsis. Plant Physiology **109**:1227–1230.

Babikova, Z., L. Gilbert, T. J. A. Bruce, M. Birkett, J. C. Caulfield, C. Woodcock, J. A. Pickett, and D. Johnson. 2013. Underground signals carried through common mycelial networks warn neighboring plants of aphid attack. Ecology Letters **16**:835–843.

Bagchi, R., T. Swinfield, R. E. Gallery, O. T. Lewis, S. Gripenberg, L. Narayan, and R. P. Freckleton. 2010. Testing the Janzen-Connell mechanism: pathogens cause overcompensating density dependence in a tropical tree. Ecology Letters **13**:1262–1269.

Bailey, S. F., A. L. Hargreaves, S. D. Hechtenthal, R. A. Laird, T. M. Latty, T. G. Reid, A. C. Teucher, and J. R. Tindall. 2007. Empty flowers as a pollination-enhancement strategy. Evolutionary Ecology Research **9**:1245–1262.

Bais, H. P., R. Vepachedu, S. Gilroy, R. M. Callaway, and J. M. Vivianco. 2003. Allelopathy and exotic plant invasion: from molecules and genes to species interactions. Science **301**:1377–1380.

Baker, H. G., and I. Baker. 1975. Studies of nectar-constitution and pollinator-plant coevolution. Pages 100–140 in L. E. Gilbert and P. H. Raven, editors. Coevolution of Plants and Animals. University of Texas Press, Austin.

Baldwin, I. T. 1998. Jasmonate-induced responses are costly but benefit plants under attack in native populations. Proceedings of the National Academy of Sciences **95**:8113–8118.

Baldwin, I. T., and J. C. Schultz. 1983. Rapid changes in tree leaf chemistry induced by damage: evidence for communication between plants. Science **221**:277–279.

Ballare, C. L. 1999. Keeping up with the neighbours: phytochrome sensing and other signalling mechanisms. Trends in Plant Science **4**:97–102.

Ballare, C. L. 2009. Illuminated behaviour: phytochrome as a key regulator of light foraging and plant anti-herbivore defence. Plant, Cell and Environment **32**:713–725.

Ballare, C. L. 2011. Jasmonate-induced defenses: a tale of intelligence, collaborators and rascals. Trends in Plant Science **16**:249–257.

Ballare, C. L., C. A. Mazza, A. T. Austin, and R. Pierik. 2012. Canopy light and plant health. Plant Physiology **160**:145–155.

Ballare, C. L., A. L. Scopel, and R. A. Sanchez. 1990. Far-red radiation reflected from adjacent leaves: an early signal of competition in plant canopies. Science **247**:329–332.

Barbosa, P., J. Hines, I. Kaplan, H. Martinson, C. Szczepaniec, and Z. Szendrei. 2009. Associational resistance and associational susceptibility: having right or wrong neighbors. Annual Review of Ecology and Systematics **40**:1–20.

Barto, E. K., M. Hilker, F. Mueller, B. K. Mohney, J. D. Weidenhamer, and M. C. Rillig. 2011. The fungal fast lane: common mycorrhizal networks extend bioactive zones of allelochemicals in soils. PLoS One **6**:e27195.

Barton, K. E., and J. Koricheva. 2010. The ontogeny of plant defense and herbivory: characterizing general patterns using meta-analysis. American Naturalist **175**:481–493.

Bate, N. J., and S. J. Rothstein. 1998. C-6-volatiles derived from the lipoxygenase pathway induce a subset of defense-related genes. Plant Journal **16**:561–569.

Beadle, G. W. 1980. The ancestry of corn. Scientific American **242**:112–119.

Beale, M. H., M. A. Birkett, T. J. A. Bruce, K. Chamberlain, L. M. Field, A. K. Huttly, J. L. Martin, R. Parker, A. L. Phillips, J. A. Pickett, I. M. Prosser, P. R. Shewry, L. E. Smart, L. J. Wadhams, C. M. Woodcock, and Y. Zhang. 2006. Aphid alarm pheromone produced by transgenic plants affects aphid and parasitoid behavior. Proceedings of the National Academy of Sciences **103**:10509–10513.

Beattie, A. J. 1985. The evolutionary ecology of ant/plant mutualism. Cambridge University Press, Cambridge.

Beatty, P. H., and A. G. Good. 2011. Future prospects for cereals that fix nitrogen. Science **333**:416–417.

Beckers, G. J. M., M. Jaskiewicz, Y. Liu, W. R. Underwood, S. Yang, S. Zhang, and U. Conrath. 2009. Mitogen-activated protein kinases 3 and 6 are required for full priming of stress responses in *Arabidopsis thaliana*. Plant Cell **21**:944–953.

Beilby, M. J. 2007. Action potential in charophytes. Pages 43–82 *in* K. W. Jeon, editor. International Review of Cytology. Elsevier, Amsterdam.

Bell, D., and L. F. Galloway. 2008. Population differentiation for plasticity to light in an annual herb: adaptation and cost. American Journal of Botany **95**:59–65.

Bell, G. 1986. The evolution of empty flowers. Journal of Theoretical Biology **118**:253–258.

Benson, W. W., K. S. Brown, and L. E. Gilbert. 1975. Coevolution of plants and herbivores—passion flower butterflies. Evolution **29**:659–680.

Bent, A. F., and D. Mackey. 2007. Elicitors, effectors, and R genes: the new paradigm and a lifetime supply of questions. Annual Review of Plant Pathology **45**:399–436.

Bentley, B. L. 1977. Extrafloral nectaries and protection by pugnacious bodyguards. Annual Review of Ecology and Systematics **8**:407–427.

Benyus, J. M. 1997. Biomimicry. Morrow, New York.

Berg, R. L. 1960. The ecological significance of correlation pleiades. Evolution **14**:171–180.

Beyaert, I., D. Kopke, J. Stiller, A. Hammerbacher, K. Yoneya, A. Schmidt, J. Gershenzon, and M. Hilker. 2012. Can insect egg deposition "warn" a plant of future feeding damage by herbivorous larvae? Proceedings of the Royal Society B **279**:101–108.

Biddington, N. L. 1986. The effects of mechanically induced stress in plants—a review. Plant Growth Regulation **4**:103–123.

Blair, A. C., L. A. Weston, S. J. Nissen, G. R. Brunk, and R. A. Hufbauer. 2009. The importance of analytical techniques in allelopathy studies with the reported allelochemical catechin as an example. Biological Invasions **11**:325–332.

Blande, J. D., J. K. Holopainen, and T. Li. 2010. Air pollution impedes plant-to-plant communication by volatiles. Ecology Letters **13**:1172–1181.

Blande, J. D., T. Li, and J. K. Holopainen. 2011. Air pollution impedes plant-to-plant communication, but what is the signal? Plant Signaling and Behavior **6**:1016–1018.

Bleecker, A. B., M. A. Estelle, C. Somerville, and H. Kende. 1988. Insensitivity to ethylene conferred by a dominant mutation in *Arabidopsis thaliana*. Science **241**:1086–1089.

Blodner, C., C. Goebel, I. Feussner, C. Gatz, and A. Polle. 2007. Warm and cold parental reproductive environments affect seed properties, fitness, and cold responsiveness in *Arabidopsis thaliana* progenies. Plant, Cell and Environment **30**:165–175.

Bloom, A. J., F. S. Chapin, and H. A. Mooney. 1985. Resource limitation in plants—an economic analogy. Annual Review of Ecology and Systematics **16**:363–392.

Bolhuis, J. J. 1991. Mechanisms of avian imprinting—a review. Biological Reviews of the Cambridge Philosophical Society **66**:303–345.

Bonaventure, G., A. Gfeller, W. M. Proebsting, S. Hortensteiner, A. Chetelat, E. Martinoia, and E. E. Farmer. 2007. A gain-of-function allele of TPC1 activates oxylipin biogenesis after leaf wounding in Arabidopsis. Plant Journal **49**:889–898.

Bonduriansky, R., and T. Day. 2009. Nongenetic inheritance and its evolutionary implications. Annual Review of Ecology and Systematics **40**:103–125.

Bostock, R. M. 2005. Signal crosstalk and induced resistance: straddling the line between cost and benefit. Annual Review of Phytopathology **43**:545–580.

Bouche, N., and H. Fromm. 2004. GABA in plants: just a metabolite? Trends in Plant Science **9**:110–115.

Bown, A. W., D. E. Hall, and K. B. MacGregor. 2002. Insect footsteps on leaves stimulate the accumulation of 4-aminobutyrate and can be visualized through increased chlorophyll fluorescence and superoxide production. Plant Physiology **129**:1430–1434.

Boyer, J. S. 1970. Leaf enlargement and metabolic rates in corn, soybean, and sunflower at various leaf water potentials. Plant Physiology **46**:233–235.

Boyko, A., D. Hudson, P. Bromkar, P. Kathiria, and I. Kovalchuk. 2006. Increase of homologous recombination frequency in vascular tissue of *Arabidopsis* plants exposed to salt stress. Plant and Cell Physiology **47**:736–742.

Braam, J. 2005. In touch: plant responses to mechanical stimuli. New Phytologist **165**:373–389.

Bradbury, J. W., and S. L. Vehrencamp. 1998. Principles of Animal Communication. Sinauer, Sunderland, MA.

Bradford, K. J. 1986. Manipulation of seed water relations via osmotic priming to improve germination under stress conditions. HortScience **21**:1105–1112.

Bradshaw, A. D. 1965. Evolutionary significance of phenotypic plasticity in plants. Advances in Genetics **13**:115–155.

Brady, C. J. 1987. Fruit ripening. Annual Review of Plant Physiology **38**:155–178.

Breed, M. D., and J. Moore. 2011. Animal Behavior. Academic Press, Burlington, MA.

Brewbaker, J., and S. K. Majumder. 1961. Cultural studies of pollen population effect and self-incompatibility inhibition. American Journal of Botany **48**:457–464.

Briggs, W. R., and J. M. Christie. 2002. Phototropins 1 and 2: versatile plant blue-light receptors. Trends in Plant Science **7**:204–210.

Briggs, W. R., and C. T. Lin. 2012. Photomorphogenesis—from one photoreceptor to 14: 40 years of progress. Molecular Plant **5**:531–532.

Brody, A. K., and R. J. Mitchell. 1997. Effects of experimental manipulation of inflorescence size on pollination and pre-dispersal seed predation in the hummingbird-pollinated plant *Ipomopsis aggregata*. Oecologia **110**:85–93.

Bronmark, C. 1989. Interactions between epiphytes, macrophytes and fresh-water snails—a review. Journal of Molluscan Studies **55**:299–311.

Brouwer, R. 1963. Some aspects of the equilibrium between overground and underground plant parts. Jaarboek van het Instituut voor Biologisch en Scheikundig onderzoek aan Landbouwgewassen **1963**:31–39.

Brown, J. H., and A. Kodric-Brown. 1979. Convergence, competition, and mimicry in a temperate community of hummingbird pollinated flowers. Ecology **60**:1022–1035.

Bruce, T. J. A. 2010. Exploiting plant signals in sustainable agriculture. Pages 215–227 in F. Baluska and V. Ninkovic, editors. Plant Communication from an Ecological Perspective. Springer-Verlag, Berlin.

Bruce, T. J. A., M. C. Matthes, K. Chamberlain, C. M. Woodcock, A. Mohib, B. Webster, L. E. Smart, M. A. Birkett, J. A. Pickett, and J. A. Napier. 2008. *cis*-Jasmone induces Arabidopsis genes that affect the chemical ecology of multitrophic interactions with aphids and their parasitoids. Proceedings of the National Academy of Sciences **105**:4553-4558.

Bruce, T. J. A., M. C. Matthes, J. A. Napier, and J. A. Pickett. 2007. Stressful "memories" of plants: evidence and possible mechanisms. Plant Science **173**:603–608.

Bruce, T. J. A., and J. A. Pickett. 2011. Perception of plant volatile blends by herbivorous insects—finding the right mix. Phytochemistry **72**:1605–1611.

Bruce, T. J. A., L. J. Wadhams, and C. M. Woodcock. 2005. Insect host location: a volatile situation. Trends in Plant Science **10**:269–274.

Brundrett, M. C. 1991. Mycorrhizas in natural ecosystems. Advances in Ecological Research **21**:171–313.

Burkle, L., and R. Irwin. 2009. Nectar sugar limits larval growth of solitary bees (Hymenoptera: Megachilidae). Environmental Entomology **38**:1293–1300.

Burkle, L. A., R. E. Irwin, and D. A. Newman. 2007. Predicting the effects of nectar robbing on plant reproduction: implications of pollen limitation and plant mating system. American Journal of Botany **94**:1935–1943.

Burrill, T. J., and R. Hansen. 1917. Is symbiosis possible between legume bacteria and non-legume plants? Bulletin of the Illinois Agricultural Experiment Station **202**:111–181.

Cahill, J. F., T. Bao, M. Maloney, and C. Kolenosky. 2013. Mechanical leaf damage causes localized, but not systemic, changes in leaf movement behavior of the "sensitive plant," *Mimosa pudica* (Fabaceae) L. Botany **9**:43-47.

Cahill, J. F., and G. G. McNickle. 2011. The behavioral ecology of nutrient foraging by plants. Annual Review of Ecology, Evolution, and Systematics **42**:289–311.

Cahill, J. F., G. G. McNickle, J. J. Haag, E. G. Lamb, S. M. Nyanumba, and C. C. St. Clair. 2010. Plants integrate informaton about nutrients and neighbors. Science **328**:1657.

Cameron, R. K., N. L. Paiva, C. J. Lamb, and R. A. Dixon. 1999. Accumulation of salicylic acid and PR-1 gene transcripts in relation to the systemic acquired resistance (SAR) response induced by *Pseudomonas syringae* pv. tomato in Arabidopsis. Physiological and Molecular Plant Pathology **55**:121–130.

Campbell, D. R. 1996. Evolution of floral traits in a hermaphroditic plant: field measurements of heritabilities and genetic correlations. Evolution **50**:1442–1453.

Campbell, D. R., N. M. Waser, and E. J. Melendez-Ackerman. 1997. Analyzing pollinator-mediated selection in a plant hybrid zone: hummingbird visitation patterns on three spatial scales. American Naturalist **149**:295–315.

Campbell, D. R., N. M. Waser, and M. V. Price. 1994. Indirect selection of stigma position in *Ipomopsis aggregata* via a genetically correlated trait. Evolution **48**:55–68.

Carmona, D., M. J. Lejeunesse, and M. T. J. Johnson. 2011. Plant traits that predict resistance to herbivores. Functional Ecology **25**:358–367.

Cashmore, A. R., J. A. Jarillo, Y.-J. Wu, and D. Liu. 1999. Cryptochromes: blue light receptors for plants and animals. Science **284**:760–765.

Caspari, E. W., and R. E. Marshak. 1965. Rise and fall of Lysenko. Science **149**:275–278.

Casper, B. B., H. J. Schenk, and R. B. Jackson. 2003. Defining a plant's belowground zone of influence. Ecology **84**:2313–2321.

Castellanos, M. C., P. Wilson, and J. D. Thomson. 2004. "Anti-bee" and "pro-bird" changes during the evolution of hummingbird pollination in Penstamon flowers. Journal of Evolutionary Biology **17**:876–885.

Catoni, C., H. M. Schaefer, and A. Peters. 2008. Fruit for health: the effect of flavonoids on humoral immune response and food selection in a frugivorous bird. Functional Ecology **22**:649–654.

Cayuela, E., F. Perez-Alfocea, M. Caro, and M. C. Bolarin. 1996. Priming of seeds with NaCl induces physiological changes in tomato plants grown under salt stress. Physiologia Plantarum **96**:231–236.

Cazetta, E., H. M. Schaefer, and M. Galetti. 2009. Why are fruits colorful? The relative importance of achromatic and chromatic contrasts for detection by birds. Evolutionary Ecology **23**:233–244.

Cerdan, P. D., and J. Chory. 2003. Regulation of flowering time by light quality. Nature **423**:881–885.

Chae, K., and E. M. Lord. 2011. Pollen tube growth and guidance: roles of small, secreted proteins. Annals of Botany **108**:627–636.

Chapman, L. A., and D. R. Goring. 2010. Pollen-pistil interactions regulating successful fertilization in the Brassicaceae. Journal of Experimental Botany **61**:1987–1999.

Chazdon, R. L., and R. W. Pearcy. 1991. The importance of sunflecks for forest understory plants—photosynthetic machinery appears adapted to brief, unpredictable periods of radiation. BioScience **41**:760–766.

Chehab, E. W., E. Eich, and J. Braam. 2009. Thigmomorphogenesis: a complex plant response to mechano-stimulation. Journal of Experimental Botany **60**:43–56.

Chehab, E. W., C Yao, Z. Henderson, S. Kim, and J. Braam. 2012. Arabidopsis touch-induced morphogenesis is jasmonate mediated and protects against pests. Current Biology **22**:701–706.

Chen, Y.-F., N. Etheridge, and G. E. Schaller. 2005. Ethylene signal transduction. Annals of Botany **95**:901–905.

Chen, Y.-F., Z. Gao, R. J. Kerris, W. Wang, B. M. Binder, and G. E. Schaller. 2010. Ethylene receptors function as components of high-molecular-mass protein complexes in Arabidopsis. PLoS One **5**:e8640.

Chester, K. S. 1933. The problem of acquired physiological immunity in plants. Quarterly Review of Biology **8**:129–154, 275–324.

Chini, A., S. Fonseca, G. Fernandex, B. Adie, J. M. Chico, O. Lorenzo, G. Garcia-Casado, I. Lopez-Vidriero, F. M. Lozano, M. R. Ponce, J. L. Micol, and R. Solano. 2007. The JAZ family of repressors is the missing link in jasmonate signalling. Nature **448**:666–671.

Chinnusamy, V., J.-K. Zhu, and R. Sunkar. 2010. Gene regulation during cold stress acclimation in plants. Pages 39–55 *in* R. Sunkar, editor. Plant Stress Tolerance. Methods in Molecular Biology **639**:39–55.

Chinnusamy, V., J. Zhu, and J.-K. Zhu. 2007. Cold stress regulation of gene expression in plants. Trends in Plant Science **12**:444–451.

Chittka, L., J. D. Thomson, and N. M. Waser. 1999. Flower constancy, insect psychology, and plant evolution. Naturwissenschaften **86**:361–377.

Choh, Y., T. Shimoda, R. Ozawa, M. Dicke, and J. Takabayashi. 2004. Exposure of lima bean leaves to volatiles from herbivore-induced conspecific plants results in emission of carnivore attractants: active or passive process? Journal of Chemical Ecology **30**:1305–1317.

Choh, Y., and J. Takabayashi. 2006. Herbivore-induced extrafloral nectar production in lima bean plants enhanced by previous exposure to volatiles from infested conspecifics. Journal of Chemical Ecology **32**:2073–2077.

Chong, J., M.-A. Pierrel, R. Atanassova, D. Werck-Reichhart, B. Fritig, and P. Saindrenan. 2001. Free and conjugated benzoic acid in tobacco plants and cell cultures: induced accumulation upon elicitation of defense responses and role as salicylic acid precursors. Plant Physiology **125**:318–328.

Christie, J. M., and A. S. Murphy. 2013. Shoot phototropism in higher plants: new light through old concepts. American Journal of Botany **100**:35–46.

Cipollini, D. F. 2002. Does competition magnify the fitness costs of induced responses in *Arabidopsis thaliana*? A manipulative approach. Oecologia **131**:514–520.

Cipollini, M. L., and D. J. Levey. 1991. Why some fruits are green when they are ripe: carbon balance in fleshy fruits. Oecologia **88**:371–377.

Cipollini, M. L., and D. J. Levey. 1997a. Why are some fruits toxic? Glycoalkaloids in *Solanum* and fruit choice by vertebrates. Ecology **78**:782–798.

Cipollini, M. L., and D. J. Levey. 1997b. Secondary metabolites of fleshy vertebrate-dispersed fruits: adaptive hypotheses and implications for seed dispersal. American Naturalist **150**:346–372.

Clarkson, D. T. 1985. Factors affecting mineral nutrient acquisition by plants. Annual Review of Plant Physiology and Plant Molecular Biology **36**:77–115.

Coleman, J. S., and C. G. Jones. 1991. A phytocentric perspective of phytochemical induction by herbivores. Pages 3–45 *in* D. W. Tallamy and M. J. Raupp, editors. Phytochemical Induction by Herbivores. John Wiley, New York.

Coleman, R. A., A. M. Barker, and M. Fenner. 1999. Parasitism of the herbivore *Pieris brassicae* L. (Lep., Pieridae) by *Cotesia glomerata* L. (Hym., Braconidae) does not benefit the host plant by reduction of herbivory. Journal of Applied Entomology **123**:171–177.

Connell, J. H. 1971. On the role of natural enemies in preventing competitive exclusion in some marine animals and in rain forest trees. Pages 298–312 *in* P. J. den Boer and G. Gradwell, editors. Dynamics of Populations. Centre for Agricultural Publishing and Documentation, Wageningen.

Conner, J. K. 2006. Ecological genetics of floral evolution. Pages 260–277 *in* L. D. Harder and S. C. H. Barrett, editors. Ecology and Evolution of Flowers. Oxford University Press, Oxford.

Conner, J. K., and A. Sterling. 1995. Testing hypotheses of functional relationships: a comparative survey of correlation patterns among floral traits in five insect-pollinated plants. American Journal of Botany **82**:1399–1406.

Conrath, U. 2009. Priming of induced plant defense responses. Advances in Botanical Research **51**:361–395.

Conrath, U., G. J. M. Beckers, V. Flors, P. Garcia-Agustin, G. Jakab, F. Mauch, M.-A. Newman, C. M. J. Pieterse, B. Poinssot, M. J. Pozo, A. Pugin, U. Schaffrath, J. Ton, D. Wendehenne, L. Zimmerli, and B. Mauch-Mani. 2006. Priming: getting ready for battle. Molecular Plant-Microbe Interactions **19**:1062–1071.

Cook, A. D., P. R. Atsatt, and C. A. Simon. 1971. Doves and dove weed: multiple defenses against avian predation. BioScience **21**:277–281.

Cook, S. M., Z. R. Khan, and J. A. Pickett. 2007. The use of push-pull strategies in integrated pest management. Annual Review of Entomology **52**:375–400.

Cornelissen, T., G. W. Fernandes, and M. S. Coelho. 2011. Induced responses in the neotropical shrub Bauhinia brevipes Vogel: does early season herbivory function as cue to plant resistance? Arthropod-Plant Interactions **5**:245–253.

Creelman, R. A., and J. E. Mullet. 1997. Biosynthesis and action of jasmonates in plants. Annual Review of Plant Physiology and Plant Molecular Biology **48**:355–381.

Cronin, T. W., and N. J. Marshall. 1989. A retina with at least ten spectral types of photoreceptors in a mantis shrimp. Nature **339**:137–140.

Cruzan, M. B. 1990. Variation in pollen size, fertilization ability, and postfertilization siring ability in Erythronium garniflorum. Evolution **44**:843–856.

Culver, D. C., and A. J. Beattie. 1980. The fate of Viola seeds dispersed by ants. American Journal of Botany **67**:710–714.

Cunningham, J. P., C. J. Moore, M. P. Zalucki, and B. W. Cribb. 2006. Insect odour perception: recognition, of odour components by flower foraging moths. Proceedings of the Royal Society B **273**:2035–2040.

Cunningham, J. P., C. J. Moore, M. P. Zalucki, and S. A. West. 2004. Learning, odour preference and flower foraging in moths. Journal of Experimental Biology **207**:87–94.

Dalin, P., and C. Bjorkman. 2003. Adult beetle grazing induced willow trichome defence against subsequent larval feeding. Oecologia **134**:112–118.

Darwin, C. 1859. On the Origin of Species by Means of Natural Selection, or the Preservation of Favoured Races in the Struggle for Life. John Murray, London.

Darwin, C. 1862. On the Various Contrivances by which British and Foreign Orchids Are Fertilized by Insects. John Murray, London.

Darwin, C. 1876. The Effects of Cross- and Self-Fertilization in the Animal Kingdom. John Murray, London.

Darwin, C. 1880. The Power of Movement in Plants. John Murray, London.

Darwin, C. 1881. Letter to Karl Semper.

Darwin, C. 1893. Insectivorous Plants. John Murray, London.

Darwin, E. 1794. Zoonomia, or the Laws of Organic Life. J. Johnson, London.

Davidovich-Rikanati, R., Y. Sitrit, Y. Tadmor, Y. Iijima, N. Bilenko, E. Bar, B. Carmona, E. Fallik, N. Dudai, J. E. Simon, E. Pichersky, and E. Lewinsohn. 2007. Enrichment of tomato flavor by diversion of the early plastidial terpenoid pathway. Nature Biotechnology **25**:899–901.

Davies, E. 2004. New functions for electrical signals in plants. New Phytologist **161**:607–610.

Davis, T. S., T. L. Crippen, R. W. Hofstetter, J. K. Tomberlin. 2013. Microbial volatile emissions as insect semiochemicals. Journal of Chemical Ecology **39**:840-859.

Dawson, T. P., S. T. Jackson, J. I. House, I. C. Prentice, and G. M. Mace. 2011. Beyond predictions: biodiversity conservation in a changing climate. Science **332**:53–58.

de Boer, J. G., T. A. L. Snoeren, and M. Dicke. 2005. Predatory mites learn to discriminate between plant volatiles induced by prey and nonprey herbivores. Animal Behaviour **69**:869–879.

de Kroon, H., H. Huber, J. F. Stuefer, and J. M. van Groenendael. 2005. A modular concept of phenotypic plasticity in plants. New Phytologist **166**:73–82.

de Kroon, H., E. J. W. Visser, H. Huber, L. Mommer, and M. J. Hutchings. 2009. A modular concept of plant foraging behaviour: the interplay between local responses and systemic control. Plant, Cell and Environment **32**:704–712.

De Moraes, C. M., W. J. Lewis, P. W. Pare, and J. H. Tumlinson. 1998. Herbivore infested plants selectively attract parasitoids. Nature **393**:570–574.

De Moraes, C. M., M. C. Mescher, and J. H. Tumlinson. 2001. Caterpillar-induced nocturnal plant volatiles repel conspecific females. Nature **410**:577–580.

de Roman, M., I. Fernandez, T. Wyatt, M. Sahrawy, M. Heil, and M. J. Pozo. 2011. Elicitation of foliar resistance mechanisms transiently impairs root association with arbuscular mycorrhizal fungi. Journal of Ecology **99**:36–45.

de Wit, M., W. Kegge, J. B. Evers, M. Vergeer-van Eijk, P. Gankema, L. A. C. J. Voesenek, and R. Pierik. 2012. Plant neighbor detection through touching leaf tips precedes phytochrome signals. Proceedings of the National Academy of Sciences **109**:14705–14710.

de Wit, M., S. H. Spoel, G. F. Sanchez-Perez, C. M. M. Gommers, C. M. J. Pieterse, L. A. C. J. Voesenek, and R. Pierik. 2013. Perception of low red:far-red ratio compromises both salicylic acid– and jasmonic acid–dependent pathogen defences in Arabidopsis. Plant Journal **75**:90–103.

Degenhardt, J., I. Hiltpold, T. G. Koellner, M. Frey, A. Gierl, J. Gershenzon, B. E. Hibbard, M. R. Ellersiecke, and T. C. J. Turlings. 2009. Restoring a maize root signal that attracts insect-killing nematodes to control a major pest. Proceedings of the National Academy of Sciences **106**:13213–13218.

Del Guidice, L., D. R. Massardo, P. Pontieri, C. M. Bertea, D. Mombello, E. Carata, S. M. Tredici, A. Tala, M. Mucciarelli, V. I. Groudeva, M. De Stefano, G. Vigliotta, M. E. Maffei, and P. Alifano. 2008. The microbial community of Vetiver root and its involvement into essential oil biogenesis. Environmental Microbiology **10**:2824–2841.

Delledonne, M. 2005. NO news is good news for plants. Current Opinion in Plant Biology **8**:390–396.

Denison, R. F. 2012. Darwinian Agriculture: How Understanding Evolution Can Improve Agriculture. Princeton University Press, Princeton, NJ.

Denison, R. F., E. T. Kiers, and S. A. West. 2003. Darwinian agriculture: when can humans find solutions beyond the reach of natural selection? Quarterly Review of Biology **78**:145–168.

DeWitt, T. J., A. Sih, and D. S. Wilson. 1998. Costs and limits of phenotypic plasticity. Trends in Ecology and Evolution **13**:77–81.

Dicke, M. 1986. Volatile spider-mite pheromone and host-plant kairomone, involved in spaced-out gregariousness in the spider mite *Tetranychus urticae*. Physiological Entomology **11**:251–262.

Dicke, M. 1999. Are herbivore-induced plant volatiles reliable indicators of herbivore identity to foraging carnivorous arthropods? Entomologia Experimentalis et Applicata **91**:131–142.

Dicke, M., M. W. Sabelis, J. Takabayashi, J. Bruin, and M. A. Posthumus. 1990. Plant strategies of manipulating predator-prey interactions through allelochemicals: prospects for application in pest control. Journal of Chemical Ecology **16**:3091–3117.

Dicke, M., R. Gols, D. Ludeking, and M. A. Posthumus. 1999. Jasmonic acid and herbivory differentially induce carnivore-attracting plant volatiles in lima bean plants. Journal of Chemical Ecology **25**:1907–1922.

Dicke, M., and M. W. Sabelis. 1988. How plants obtain predatory mites as bodyguards. Netherlands Journal of Zoology **38**:148–165.

Dicke, M., P. van Baarlen, R. Wessels, and H. Dijkman. 1993. Herbivory induces systemic production of plant volatiles that attract predators of the herbivore: extraction of endogenous elicitor. Journal of Chemical Ecology **19**:581–599.

Dicke, M., and J. J. A. van Loon. 2000. Multitrophic effects of herbivore-induced plant volatiles in an evolutionary context. Entomologia Experimentalis et Applicata **97**:237–249.

Dicke, M., and R. M. P. van Poecke. 2002. Signalling plant-insect interactions: signal transduction in direct and indirect plant defence. Pages 289–316 *in* D. Scheel and C. Wasternack, editors. Frontiers in Molecular Biology: Plant Signal Transduction. Oxford University Press, Oxford.

Doares, S. H., J. Narvaez-Vasquez, A. Conconi, and C. A. Ryan. 1995. Salicylic-acid inhibits synthesis of proteinase-inhibitors in tomato leaves induced by systemin and jasmonic acid. Plant Physiology **108**:1741–1746.

Dodd, M. E., J. Silvertown, and M. W. Chase. 1999. Phylogenetic analysis of trait evolution and species diversity variation among angiosperm families. Evolution **53**:732–744.

Dodson, C. H., R. L. Dressler, H. G. Hills, R. M. Adams, and N. H. Williams. 1969. Biologically active compounds in orchid fragrances. Science **164**:1243–1249.

Doebley, J. 2004. The genetics of maize evolution. Annual Review of Genetics **38**:37–59.

Doherty, H. M., R. R. Selvendran, and D. J. Bowles. 1988. The wound response of tomato plants can be inhibited by aspirin and related hydroxybenzoic acids. Physiological and Molecular Plant Pathology **33**:377–384.

Dolch, R., and T. Tscharntke. 2000. Defoliation of alders (*Alnus glutinosa*) affects herbivory by leaf beetles on undamaged neighbors. Oecologia **125**:504–511.

Dominy, N. J. 2004. Fruits, fingers, and fermentation: the sensory cues available to foraging primates. Integrative and Comparative Biology **44**:295–303.

Donohue, K., E. H. Pyle, D. Messiqua, M. S. Heschel, and J. Schmitt. 2000. Density dependence and population differentiation of genetic architecture in *Impatiens capensis* in natural environments. Evolution **54**:1969–1981.

Donohue, K., E. H. Pyle, D. Messiqua, M. S. Heschel, and J. Schmitt. 2001. Adaptive divergence in plasticity in natural populations of *Impatiens capensis* and its consequences for performance in novel habitats. Evolution **55**:692–702.

Doss, R. P., J. E. Oliver, W. M. Proebsting, S. W. Potter, S. R. Kuy, S. L. Clement, R. T. Williamson, J. R. Carney, and E. D. DeVilbiss. 2000. Bruchins: insect-derived plant regulators that stimulate neoplasm formation. Proceedings of the National Academy of Sciences **97**:6218–6223.

Dowell, R. D., O. Ryan, A. Jansen, D. Cheung, S. Agarwala, T. Danford, D. A. Bernstein, P. A. Rolfe, L. E. Heisler, B. Chin, C. Nislow, G. Giaever, P. C. Phillips, G. R. Fink, D. K. Gifford, and C. Boone. 2010. Genotype to phenotype: a complex problem. Science **328**:469.

Dressler, R. L. 1982. Biology of the orchid bees (Euglossini). Annual Review of Ecology and Systematics **13**:373–394.

Drew, M. C. 1975. Comparison of the effects of a localized supply of phosphate, nitrate, ammonium and potassium on the growth of the seminal root system, and the shoot, of barley. New Phytologist **111**:479–490.

Drew, M. C., L. R. Saker, and T. W. Ashley. 1973. Nutrient supply and the growth of the seminal root system in barley. I. The effect of nitrate concentration on the growth of axes and laterals. Journal of Experimental Botany **24**:1189–1202.

Drukker, B., P. Scutareanu, and M. W. Sabelis. 1995. Do Anthocorid predators respond to synomones from Psylla-infested pear trees under field conditions? Entomologia Experimentalis et Applicata **77**:193–203.

Du, L., G. S. Ali, K. A. Simons, J. Hou, T. Yang, A. S. N. Reddy, and B. W. Poovaiah. 2009. Ca^{2+}/calmodulin regulates salicylic-acid-mediated plant immunity. Nature **457**:1154–1158.

Dudareva, N., F. Negre, D. A. Nagegowda, and I. Orlova. 2006. Plant volatiles: recent advances and future perspectives. Critical Reviews in Plant Sciences **25**:417–440.

Dudareva, N., and E. Pichersky. 2008. Metabolic engineering of plant volatiles. Current Opinion in Biotechnology **19**:181–189.

Dudareva, N., E. Pichersky, and J. Gershenzon. 2004. Biochemistry of plant volatiles. Plant Physiology **135**:1893–1902.

Dudley, S. A., and A. L. File. 2007. Kin recognition in an annual plant. Biology Letters **3**:435–438.

Dudley, S. A., and J. Schmitt. 1996. Testing the adaptive plasticity hypothesis: density-dependent selection on manipulated stem length in *Impatiens capensis*. American Naturalist **147**:445–465.

Duffey, S. S., and G. W. Felton. 1989. Plant enzymes in resistance to insects. Pages 289–313 *in* J. R. Whitaker and P. E. Sonnet, editors. Biocatalysis in Agricultural Biotechnology. American Chemical Society, Toronto.

Dukas, R., and J. J. Duan. 2000. Potential fitness consequences of associative learning in a parasitoid wasp. Behavioral Ecology **11**:536–543.

Dukas, R., and L. A. Real. 1993. Learning constaints and floral choice behavior in bumblebees. Animal Behaviour **46**:637–644.

Durrant, W. E., and X. Dong. 2004. Systemic acquired resistance. Annual Review of Phytopathology **42**:185–209.

Dyer, L. A., C. D. Dodson, J. Belhoffer, and D. K. Letourneau. 2001. Trade-offs in anti-herbivore defenses in *Piper cenocladum*: ant mutualists versus plant secondary metabolites. Journal of Chemical Ecology **27**:581–592.

Edwards, P. J., and S. D. Wratten. 1983. Wound induced defenses in plants and their consequences for patterns of insect grazing. Oecologia **59**:88–93.

Efferth, T. 2009. Artemisinin: a versatile weapon from traditional Chinese medicine. Pages 173–194 *in* K. G. Ramawat, editor. Herbal drugs: ethnomedicine to modern medicine. Springer, Berlin.

Ehrlen, J., and O. Eriksson. 1993. Toxicity in fleshly fruits: a non-adaptive trait? Oikos **66**:107–113.

Ehrlich, P. R., and P. H. Raven. 1964. Butterflies and plants: a study in coevolution. Evolution **18**:586–608.

Eisikowitch, D., and Z. Lazar. 1987. Flower change in *Oenothera drummondii* Hooker as a response to pollinators visits. Botanical Journal of the Linnean Society **95**:101–111.

Endler, J. A. 1986. Natural Selection in the Wild. Princeton University Press, Princeton, NJ.

Engleberth, J., H. T. Alborn, E. A. Schmelz, and J. H. Tumlinson. 2004. Airborne signals prime plants against herbivore attack. Proceedings of the National Academy of Sciences **101**:1781–1785.

Engleberth, J., I. Seidl-Adams, J. C. Schultz, and J. H. Tumlinson. 2007. Insect elicitors and exposure to green leafy volatiles differentially upregulate major octadecanoids and transcripts of 12-oxo phytodienoic acid reductases in Zea mays. Molecular Plant-Microbe Interactions **20**:707–716.

English-Loeb, G. M., R. Karban, and D. Hougen-Eitzman. 1993. Direct and indirect competition between spider mites feeding on grapes. Ecological Applications **3**:699–707.

Erez, A. 1987. Use of the rest avoidance technique in peaches in Israel. Acta Horticulturae **199**:137–144.

Erez, A., R. M. Samish, and S. Lavee. 1966. Role of light in leaf and flower bud break of peach (Prunus persica). Physiologia Plantarum **19**:650–659.

Faegri, K., and L. van der Pijl. 1979. The Principles of Pollination Ecology, 3rd ed. Pergamon Press, Oxford.

Falik, O., P. Reides, M. Gersani, and A. Novoplansky. 2003. Self/non-self discrimination in roots. Journal of Ecology **91**:525–531.

Falik, O., P. Reides, M. Gersani, and A. Novoplansky. 2005. Root navigation by self inhibition. Plant, Cell and Environment **28**:562–569.

Falkenstein, E., B. Groth, A. Mithoefer, and E. W. Weiler. 1991. Methyl jasmonate and alpha linolenid acid are potent inducers of tendril coiling. Planta **185**:316-322.

Farag, M. A., H. Zhang, and C.-M. Ryu. 2013. Dynamic chemical communication between plants and bacteria through airborne signals: induced resistance by bacterial volatiles. Journal of Chemical Ecology **39**:1007–1018.

Farley, R. A., and A. Fitter. 1999. The responses of seven co-occurring woodland herbaceous perennials to localized nutrient-rich patches. Journal of Ecology **87**:849–859.

Farmer, E. E. 2001. Surface-to-air signals. Nature **411**:854–856.

Farmer, E. E., and C. A. Ryan. 1990. Interplant communication: airborne methyl jasmonate induces synthesis of proteinase inhibitors. Proceedings of the National Academy of Sciences **87**:7713–7716.

Fatouros, N. E., C. Broekgaarden, G. Bukovinszkine'Kiss, J. J. A. van Loon, R. Mumm, M. E. Huigens, M. Dicke, and M. Hilker. 2008. Male-derived butterfly antiaphrodisiac mediates induced indirect plant defense. Proceedings of the National Academy of Sciences **105**:10033–10018.

Fedina, I. S., and T. D. Tsonev. 1997. Effect of pretreatment with methyl jasmonate on the response of Pisum sativum to salt stress. Journal of Plant Physiology **151**:735–740.

Fellbaum, C. R., E. W. Gachomo, Y. Beesetty, S. Choudhari, G. D. Strahan, P. E. Pfeffer, E. T. Kiers, and H. Bucking. 2012. Carbon availability triggers fungal nitrogen uptake and transport in arbuscular mycorrhizal symbiosis. Proceedings of the National Academy of Sciences **109**:2666–2671.

Felton, G. W., and J. H. Tumlinson. 2008. Plant-insect dialogs: complex interactions at the plant-insect interface. Current Opinion in Plant Biology **11**:457–463.

Felton, G. W., J. Workman, and S. S. Duffey. 1992. Avoidance of antinutritive plant defense—role of midgut pH in Colorado potato beetle. Journal of Chemical Ecology **18**:571–583.

Fenster, C. B., W. S. Armbruster, P. Wilson, M. R. Dudash, and J. D. Thomson. 2004. Pollination syndromes and floral specialization. Annual Review of Ecology and Systematics **35**:375–403.

Ferreira, J. F. S., and J. Janick. 1995. Floral morphology of *Artemisia annua* with special reference to trichomes. International Journal of Plant Sciences **156**:807–815.

Fineblum, W. L., and M. D. Rausher. 1997. Do floral pigmentation genes also influence resistance to enemies? The W locus in *Ipomoea purpurea*. Ecology **78**:1646–1654.

Finlayson, S. A., I. J. Lee, and P. W. Morgan. 1998. Phytochrome B and the regulation of circadian ethylene production in sorghum. Plant Physiology **116**:17–25.

Firestein, S. 2001. How the olfactory system makes sense of scents. Nature **413**:211–218.

Fitter, A., L. Williamson, B. Linkohr, and O. Leyser. 2002. Root system architecture determines fitness in an Arabidopsis mutant in competition for immobile phosphate ions but not for nitrate ions. Proceedings of the Royal Society B **269**:2017–1022.

Fleming, A. 1929. On the antibacterial action of cultures of a penicllium, with special reference to their use in the isolation of *B. influenzae*. British Journal of Experimental Pathology **10**:226–236.

Forde, B. G. 2002. Local and long-range signaling pathways regulating plant responses to nitrate. Annual Review of Plant Biology **53**:203–224.

Forget, P. M. 1992. Seed removal and seed fate in *Gustavia superba* (Lecythidaceae). Biotropica **24**:408–414.

Forterre, Y., J. M. Skotheim, J. Dumais, and L. Mahadevan. 2005. How the Venus flytrap snaps. Nature **433**:421–425.

Fowler, S. V., and J. H. Lawton. 1985. Rapidly induced defenses and talking trees—the devil's advocate position. American Naturalist **126**:181–195.

Fraenkel, G. S. 1959. The raison d'etre of secondary plant substances. Science **129**:1466–1470.

Franche, C., K. Lindstrom, and C. Elmerich. 2009. Nitrogen-fixing bacteria associated with leguminous and non-leguminous plants. Plant and Soil **321**:35–59.

Fransen, B., J. Blijjenberg, and H. de Kroon. 1999. Root morphological and physiological plasticity of perennial grass species and the exploitation of spatial and temporal heterogeneous nutrient patches. Plant and Soil **211**:179–189.

Fricke, E. C., M. J. Simon, K. M. Reagan, D. J. Levey, J. A. Riffell, T. A. Carlo, and J. J. Tewksbury. 2013. When condition trumps location: seed consumption by fruit-eating birds removes pathogens and predator attractants. Ecology Letters **16**:1031–1036.

Friend, A. L., M. R. Eide, and T. M. Hinckley. 1990. Nitrogen stress alters root proliferation in Douglas-fir seedlings. Canadian Journal of Forest Research **20**:1524–1529.

Fritzsche Hoballah, M. E., and T. C. J. Turlings. 2001. Experimental evidence that plants under caterpillar attack may benefit from attracting parasitoids. Evolutionary Ecology Research **3**:553–565.

Fromm, J., and S. Lautner. 2007. Electrical signals and their physiological significance in plants. Plant, Cell and Environment **30**:249–257.

Frost, C. J., M. C. Mescher, C. Dervinis, J. M. Davis, J. E. Carlson, and C. M. De Moraes. 2008. Priming defense genes and metabolites in hybrid poplar by the green leaf volatile *cis*-3-hexenyl acetate. New Phytologist **180**:722–734.

Fujimoto, R., and T. Nishio. 2007. Self-incompatibility. Advances in Botanical Research **45**:139–154.

Gaffney, T., L. Friedrich, B. Vernooij, N. D., G. Nye, S. Uknes, E. Ward, H. Kessmann, and J. A. Ryals. 1993. Requirement of salicylic acid for the induction of systemic acquired resistance. Science **261**:754–756.

Gagliano, M. 2013. Green symphonies: a call for studies on sound communication in plants. Behavioral Ecology **24**:789-796.

Gagliano, M., S. Mancuso, and D. Robert. 2012b. Towards understanding plant bio-accoustics. Trends in Plant Science **17**:323–325.

Gagliano, M., M. Renton, N. Duvdevani, M. Timmins, and S. Mancuso. 2012a. Out of sight but not out of mind: alternative means of communication in plants. PLoS One **7**:e37382.

Galen, C. 1996. Rates of floral evolution: adaptation to bumblebee pollination in an alpine wildflower, *Polemonium viscosum*. Evolution **50**:120–125.

Galen, C. 1999. Why do flowers vary? The functional ecology of variation in flower size and form within natural plant populations. BioScience **49**:631–640.

Galen, C., and J. C. Geib. 2007. Density-dependent effects of ants on selection for bumble bee pollination in *Polemonium viscosum*. Ecology **88**:1202–1209.

Galizia, C. G., J. Kunze, A. Gumbert, A.-K. Borg-Karlson, S. Sachse, C. Markl, and R. Menzel. 2005. Relationship of visual and olfactory signal parameters in a food-deceptive flower mimicry system. Behavioral Ecology **16**:159–168.

Galloway, L. F., and J. R. Etterson. 2007. Transgenerational plasticity is adaptive in the wild. Science **318**:1134–1136.

Gan, Y. B., S. Filleur, and A. Rahman. 2005. Nutritional regulation of ANR1 and other root-expressed MADS-box genes in *Arabidopsis thaliana*. Planta **222**:730–742.

Gardner, A., and S. A. West. 2006. Spite. Current Biology **16**:662–664.

Garg, A. K., R. J. H. Sawers, H. Wang, J.-K. Kim, J. M. Walker, T. P. Brutnell, M. V. Partha-sarathy, R. D. Vierstra, and R. J. Wu. 2006. Light-regulated overexpression of an Arabidopsis phytochrome A gene in rice alters plant architecture and increases grain yield. Planta **223**:627–636.

Geber, M. A., and L. R. Griffen. 2003. Inheritance and natural selection on functional traits. International Journal of Plant Sciences **164**:S21–S42.

Geervliet, J. B. F., S. Ariens, M. Dicke, and L. E. M. Vet. 1998a. Long-distance assessment of patch profitability through volatile infochemicals by the parasitoids *Cotesia glomerata* and *C. rubecula* (Hymenoptera : Braconidae). Biological Control **11**:113–121.

Geervliet, J. B. F., A. I. Vreugdenhil, M. Dicke, and L. E. M. Vet. 1998b. Learning to discriminate between infochemicals from different plant-host complexes by the parasitoids *Cotesia glomerata* and *C. rubecula*. Entomologia Experimentalis et Applicata **86**:241–252.

Gegear, R. J., and T. M. Laverty. 2005. Flower constancy in bumblebees: a test of the trait variability hypothesis. Animal Behaviour **69**:939–949.

Geiger, D., S. Scherzer, P. Mumm, I. Marten, P. Ache, S. Maschi, A. Liese, C. Wellmann, K. A. S. Al-Rasheid, E. Grill, T. Romeis, and R. Hedrich. 2010. Guard cell anion channel SLAC1 is regulated by CDPK protein kinases with distinct Ca^{2+} affinities. Proceedings of the National Academy of Sciences **107**:8023–8028.

Gelvin, S. B. 2003. *Agrobacterium*-mediated plant transformation: the biology behind the "gene-Jockeying" tool. Microbiology and Molecular Biology Reviews **67**:16–37.

Genoud, T., A. J. Buchala, N. H. Chua, and J.-P. Metraux. 2002. Phytochrome signalling modulates the SA-perceptive pathway in Arabidopsis. Plant Journal **31**:87–95.

Genre, A., M. Chabaud, T. Timmers, P. Bonfante, and D. G. Barker. 2005. Arbuscular mycorrhizal fungi elicit a novel intracellular apparatus in *Medicago truncatula* root epidermal cells before infection. Plant Cell **17**:3489–3499.

George, A. P., R. J. Nissen, and W. B. Sherman. 1988. Overlapping double and early single cropping of low-chill peaches in Australia. Fruit Varieties Journal **43**:91–95.

Gersani, M., Z. Abramsky, and O. Falik. 1998. Density-dependent habitat selection in plants. Evolutionary Ecology **12**:223–234.

Gersani, M., J. S. Brown, E. E. O'Brien, G. M. Maina, and Z. Abramsky. 2001. Tragedy of the commons as a result of root competition. Journal of Ecology **89**:660–669.

Gersani, M., and T. Sachs. 1992. Development correlations between roots in heterogeneous environments. Plant, Cell and Environment **15**:463–469.

Gershenzon, J. 2007. Plant volatiles carry both public and private messages. Proceedings of the National Academy of Sciences **104**:5257–5258.

Gershenzon, J., and R. Croteau. 1991. Terpenoids. Pages 165–219 in G. A. Rosenthal and M. R. Berenbaum, editors. Herbivores: Their Interactions with Secondary Plant Metabolites. Academic, New York.

Gianoli, E., and F. Carrasco-Urra. 2014. Leaf mimicry in a climbing plant protects against herbivory. Current Biology **24**:984-987.

Gibbs, K. A., M. L. Urbanowski, and E. P. Greenberg. 2008. Genetic determinants of self identity and social recognition in bacteria. Science **321**:256–259.

Giladi, I. 2006. Choosing benefits or partners: a review of the evidence for the evolution of myrmecochory. Oikos **112**:481–492.

Gilbert, N. 2013. A hard look at GM crops. Nature **497**:24-26.

Giovannetti, M., C. Sbrana, A. Silvia, and L. Avio. 1996. Analysis of factors involved in fungal recognition response to host-derived signals by arbuscular mycorrhizal fungi. New Phytologist **133**:65–71.

Giurfa, M., J. Nunez, L. Chittka, and R. Menzel. 1995. Colour preferences of flower naive honeybees. Journal of Comparative Physiology A **177**:247–259.

Glazebrook, J. 2005. Contrasting mechanisms of defense against biotrophic and necrotrophic pathogens. Annual Review of Phytopathology **43**:205–227.

Goddard, M. R. 2008. Quantifying the complexities of *Saccharomyces cerevisiae*'s ecosystem engineering via fermentation. Ecology **89**:2077–2082.

Goel, D., A. K. Singh, V. Yadav, S. B. Babbar, and K. C. Bansal. 2010. Overexpression of osmotin gene confers tolerance to salt and drought stresses in transgenic tomato (*Solanum lycopersicum* L.). Protoplasma **245**:133–141.

Goh, C.-H., H. G. Nam, and Y. S. Park. 2003. Stress memory in plants: a negative regulation of stomatal response and transient induction of rd22 gene to light in abscisic acid–entrained Arabidopsis plants. Plant Journal **36**:240–255.

Gomez, S., W. van Dijk, and J. F. Stuefer. 2010. Timing of induced resistance in a clonal plant network. Plant Biology **12**:512–517.

Gomez-Gomez, L., and T. Boller. 2000. FLS2: an LRR receptor-like kinase involved in the perception of the bacterial elicitor flagellin in Arabidopsis. Molecular Cell **5**:1003–1011.

Gommers, C. M. M., E. J. W. Visser, K. R. St. Onge, L. A. C. J. Voesenek, and R. Pierik. 2013. Shade tolerance: when growing tall is not an option. Trends in Plant Science **18**:65–71.

Goodrich-Tanrikulu, M., N. E. Mahoney, and S. B. Rodriguez. 1995. The plant-growth regulator methyl jasmonate inhibits aflatoxin production by *Aspergillus flavus*. Microbiology **141**:2831-2837.

Gori, D. F. 1983. Post-pollination phenomena and adaptive floral changes. Pages 31–49 in C. E. Jones and R. J. Little, editors. Handbook of Experimental Pollination Biology. Van Nostrand Reinhold, New York.

Gorlach, J., S. Volrath, G. Knaufbeiter, G. Hengy, U. Beckhove, K. Kogel, M. Oosterdorp, T. Staub, E. Ward, H. Kessmann, and J. A. Ryals. 1996. Benzothiadiazole, a novel class of inducers of systemic acquired resistance, activates gene expression and disease resistance in wheat. Plant Cell **8**:629–643.

Gould, S. J., and R. C. Lewontin. 1979. The spandrels of San Marco and the Panglossian paradigm: a critique of the adaptationist programme. Proceedings of the Royal Society B **205**:581–598.

Goulson, D. 2000. Are insects flower constant because they use search images to find flowers? Oikos **88**:547–552.

Grant, V. 1949. Pollination systems as isolating mechanisms in angiosperms. Evolution **3**:82–97.

Grant, V., and K. A. Grant. 1965. Flower Pollination in the Phlox Family. Columbia University Press, New York.

Green, T. R., and C. A. Ryan. 1972. Wound-induced proteinase inhibitor in plant leaves: a possible defense mechanism against insects. Science **175**:776–777.

Greenup, A., W. J. Peacock, E. S. Dennis, and B. Trevaskis. 2009. The molecular biology of seasonal flowering-responses in Arabidopsis and the cereals. Annals of Botany **103**:1165–1172.

Griebel, T., and J. Zeier. 2008. Light regulation and daytime dependency of inducible plant defenses in Arabidopsis: phytochrome signaling controls systemic acquired resistance rather than local defense. Plant Physiology **147**:790–801.

Grier, J. W., and T. Burk. 1992. Biology of Animal Behavior, 2nd ed. Times Mirror/Mosby, St. Louis.

Grime, J. P. 1979. Plant Strategies and Vegetative Proceses. John Wiley, Chichester, UK.

Gronquist, M., A. Bezzerides, A. Attygalle, J. Meinwald, M. Eisner, and T. Eisner. 2001. Attractive and defensive functions of the ultraviolet pigments of a flower (*Hypericum calycinum*). Proceedings of the National Academy of Sciences **98**:13745–13750.

Gross, K. L., A. Peters, and K. S. Pregitzer. 1993. Fine root growth and demographiic responses to nutrient patches in four old-field plant species. Oecologia **95**:61–64.

Gruntman, M., and A. Novoplansky. 2004. Physiologically mediated self/non-self discrimination in roots. Proceedings of the National Academy of Sciences **101**:3863–3867.

Guy, C. L. 1990. Cold acclimation and freezing stress tolerance: role of protein metabolism. Annual Review of Plant Physiology and Plant Molecular Biology **41**:187–223.

Hadwiger, L. A. 2013. Multiple effects of chitosan on plant systems: solid science or hype. Plant Science **208**:42–49.

Hall, D. E., K. B. MacGregor, J. Nijsse, and A. W. Bown. 2004. Footsteps from insect larvae damage leaf surfaces and initiate rapid responses. European Journal of Plant Pathology **110**:441–447.

Halliday, K. J., M. Koornneef, and G. C. Whitelam. 1994. Phytochrome B and at least one other phytochrome mediate the accelerated flowering response of *Arabidopsis thaliana* L. to low red/far-red ratio. Plant Physiology **104**:1311–1315.

Hamilton, W. D. 1964. The genetical evolution of social behaviour. Journal of Theoretical Biology **7**:1–52.

Hammons, D. L., K. S. Kaan, M. C. Newman, and D. A. Potter. 2009. Invasive Japanese beetles facilitate aggregation and injury by a native scarab pest of ripening fruits. Proceedings of the National Academy of Sciences **106**:3686–3691.

Hansen, D. M., T. Van der Niet, and S. D. Johnson. 2012. Floral signposts: testing the significance of visual "nectar guides" for pollinator behaviour and plant fitness. Proceedings of the Royal Society B **279**:634–639.

Hansen, T. F., C. Pelabon, and D. Houle. 2011. Heritability is not evolvability. Evolutionary Biology **38**:258–277.

Harder, L. D., and S. C. H. Barrett. 1992. The energy-cost of bee pollination for *Pontederia cordata* (Pontederiaceae). Functional Ecology **6**:226–233.

Hare, J. D. 2011. Ecological role of volatiles produced by plants in response to damage by herbivorous insects. Annual Review of Entomology **56**:161–180.

Harms, K. E., S. J. Wright, O. Calderon, A. Hernandez, and A. E. Herre. 2000. Pervasive density-dependent recruitment enhances seedling diversity in a tropical forest. Nature **404**:493–495.

Harper, J. L. 1985. Modules, branches, and the capture of resources. Pages 1–33 *in* J. B. C. Jackson, L. W. Buss, and R. E. Cook, editors. Population Biology and Evolution of Clonal Organisms. Yale University Press, New Haven, CT.

Hartley, S. E., and J. H. Lawton. 1991. Biochemical aspects and significance of the rapidly induced accumulation of phenolics in birch foliage. Pages 105–132 *in* D. W. Tallamy and M. J. Raupp, editors. Phytochemical Induction by Herbivores. John Wiley, New York.

Hartnett, D. C., and F. A. Bazzaz. 1985. The integration of neighborhood effects by clonal genets in *Solidago canadensis*. Journal of Ecology **73**:415–427.

Harvell, C. D. 1990. The ecology and evolution of inducible defenses. Quarterly Review of Biology **65**:323–340.

Haukioja, E., and T. Hakala. 1975. Herbivore cycles and periodic outbreaks: formulation of a general hypothesis. Report from the Kevo Subarctic Research Station **12**:1–9.

Haukioja, E., and S. Neuvonen. 1985. Induced long-term resistacne of birch foliage against defoliators: defensive or incidental? Ecology **66**:1303–1308.

Haukioja, E., and P. Niemela. 1977. Retarded growth of a geometrid larva after mechanical damage to leaves of its host tree. Annales Zoololici Fennici **14**:48–52.

Hauser, M. D. 1996. The Evolution of Communication. MIT Press, Cambridge, MA.

Hayashi, T., A. Harada, T. Sakai, and S. Takagi. 2006. Ca^{2+} transient induced by extracellular changes in osmotic pressure in Arabidopsis leaves: differential involvement of cell wall-plasma membrane adhesion. Plant, Cell, and Environment **29**:661–672.

He, X. H., C. Critchley, H. Ng, and C. Bledson. 2005. Nodulated N_2-fixing *Casuarina cunninghamiana* is the sink for net N transfer from non-N_2-fixing *Eucalyptus maculata* via an ectomycorrhizal fungus *Pisolithus* sp. using $^{15}NH_4^+$ or $^{15}NO_3^-$ supplied as ammonium nitrate. New Phytologist **167**:897–912.

Heil, M. 2004. Induction of two indirect defenses benefits lima been (*Phaseolus lunatus*, Fabaceae) in nature. Journal of Ecology **92**:527–536.

Heil, M. 2008. Indirect defence via tritrophic interactions. New Phytologist **178**:41–61.

Heil, M., and R. M. Adame-Alvarez. 2010. Short signalling distances make plant communication a soliloquy. Biology Letters **6**:843–845.

Heil, M., M. Gonzalez-Teuber, L. W. Clement, S. Kautz, M. Verhaagh, and J. C. Silva Bueno. 2009. Divergent investment strategies of Acacia myrmecophytes and the coexistence of mutualists and exploiters. Proceedings of the National Academy of Sciences **106**:18091–18096.

Heil, M., and R. Karban. 2010. Explaining evolution of plant communication by airborne signals. Trends in Ecology and Evolution **25**:137–144.

Heil, M., T. Koch, A. Hilpert, B. Fiala, W. Boland, and K. E. Linsenmair. 2001. Extrafloral nectar production of the ant-associated plant, *Macaranga tanarius*, is an induced, indirect, defensive response elicited by jasmonic acid. Proceedings of the National Academy of Sciences **98**:1083–1088.

Heil, M., J. Rattke, and W. Boland. 2005. Postsecretory hydrolysis of nectar sucrose and specialization in ant/plant mutualism. Science **308**:560–563.

Heil, M., and J. C. Silva Bueno. 2007. Within-plant signaling by volatiles leads to induction and priming of an indirect plant defense in nature. Proceedings of the National Academy of Sciences **104**:5467–5472.

Heiling, A. M., M. E. Herberstein, and L. Chittka. 2003. Pollinator attraction: crab-spiders manipulate flower signals. Nature **421**:334.

Heinrich, B. 1979. Majoring and minoring by foraging bumblebees, *Bombus vagans*, experimental analysis. Ecology **60**:245–255.

Heinrich, B., and P. H. Raven. 1972. Energetics and pollination ecology. Science **176**:597–602.

Heithaus, E. R. 1981. Seed predation by rodents on three ant-dispersed plants. Ecology **62**:136–145.

Helms, A. M., C. M. De Moraes, J. F. Tooker, and M. C. Mescher. 2013. Exposure of *Solidago altissima* plants to volatile emissions of an insect antagonist (*Eurosta solidaginis*) deters subsequent herbivory. Proceedings of the National Academy of Sciences **110**:199–204.

Herrera, C. M. 1982. Defense of ripe fruit from pests: its significance in relation to plant-disperser interactions. American Naturalist **120**:218–241.

Herrera, C. M. 2001. Deconstructing a floral phenotype: do pollinators select for corolla integration in *Lavandula latifolia*? Journal of Evolutionary Biology **14**:574–584.

Herrera, C. M. 2009. Multiplicity in Unity. University of Chicago Press, Chicago.

Herrero, M., and J. Hormaza, I. 1996. Pistil strategies controlling pollen tube growth. Sexual Plant Reproduction **9**:343–347.

Hetrick, B. A. D., G. W. T. Wilson, and J. F. Leslie. 1991. Root architecture of warm-season and cool-season grasses—relationship to mycorrhizal dependence. Canadian Journal of Botany **69**:112–118.

Hilker, M., and T. Meiners. 2010. How do plants "notice" attack by herbivorous arthropods? Biological Reviews **85**:267–280.

Hinsinger, P. 2001. Bioavailability of soil inorganic P in the rhizosphere as affected by root-induced chemical changes: a review. Plant and Soil **237**:173–195.

Hodge, A. 2004. The plastic plant: root responses to heterogeneous supplies of nutrients. New Phytologist **162**:9–24.

Hodge, A. 2009. Root decisions. Plant, Cell and Environment **32**:628–640.

Hodge, A., D. Robinson, B. S. Griffiths, and A. H. Fitter. 1999. Why plants bother: root proliferation results in increased nitrogen capture from an organic patch when two grasses compete. Plant, Cell and Environment **22**:811–820.

Hoffland, E., R. Vandenboogaard, J. Nelemans, and G. Findenegg. 1992. Biosynthesis and root exudation of citric and malic-acids in phosphate-starved rape plants. New Phytologist **122**:675–680.

Hohm, T., T. Preuten, and C. Fankhauser. 2013. Phototropism: translating light into directional growth. American Journal of Botany **100**:47–59.

Holeski, L. M. 2007. Within and between generation phenotypic plasticity in trichome density of *Mimulus guttatus*. Journal of Evolutionary Biology **20**:2092–2100.

Holeski, L. M., G. Jander, and A. A. Agrawal. 2012. Transgenerational defense induction and epigenetic inheritance in plants. Trends in Ecology and Evolution **27**:618–626.

Holopainen, J. K., and J. D. Blande. 2012. Molecular plant volatile communication. Pages 17–31 *in* C. Lopez-Larrea, editor. Sensing in Nature. Landes Bioscience, Austin, Texas.

Holzapfel, C., and P. Alpert. 2003. Root cooperation in a clonal plant: connected strawberries segregate roots. Oecologia **134**:72–77.

Horiuchi, J., G. Arimura, R. Ozawa, T. Shimoda, J. Takabayashi, and T. Nishioka. 2003. A comparison of the responses of *Tetranychus urticae* (Acari : Tetranychidae) and *Phytoseiulus persimilis* (Acari : Phytoseiidae) to volatiles emitted from lima bean leaves with different levels of damage made by *T. urticae* or *Spodoptera exigua* (Lepidoptera : Noctuidae). Applied Entomology and Zoology **38**:109–116.

Hougen-Eitzman, D., and R. Karban. 1995. Mechanisms of interspecific competition that result in successful control of Pacific mites following inoculations of Willamette mites on grapevines. Oecologia **103**:157–161.

Hountondji, F. C. C., M. W. Sabelis, R. Hanna, and A. Janssen. 2005. Herbivore-induced plant volatiles trigger sporulation in entomopathogenic fungi: the case of *Neozygites tanajoae* infecting the cassava green mite. Journal of Chemical Ecology **31**:1003–1021.

Howe, G. A., and G. Jander. 2008. Plant immunity to insect herbivores. Annual Review of Plant Biology **59**:41–66.

Howe, H. F., and J. Smallwood. 1982. Ecology of seed dispersal. Annual Review of Ecology and Systematics **13**:201–228.

Huang, M., A. M. Sanchez-Moreiras, C. Abel, R. Sohrabi, S. Lee, J. Gershenzon, and D. Tholl. 2012. The major volatile organic compound emitted from *Arabidopsis thaliana* flowers, the sesquiterpene (E)-beta-caryophyllene, is a defense against a bacterial pathogen. New Phytologist **193**:997–1008.

Huber, H., N. C. Kane, M. S. Heschel, E. J. von Wettberg, J. Banta, A.-M. Leuck, and J. Schmitt. 2004. Frequency and microenvironmental pattern of selection on plastic shade-avoidance traits in a natural population of *Impatiens capensis*. American Naturalist **163**:548–563.

Hughes, L., M. Dunlop, K. French, M. R. Lieshman, B. Rice, L. Rodgerson, and M. Westoby. 1994. Predicting dispersal spectra: a minimal set of hypotheses based on plant attributes. Journal of Ecology **82**:933–950.

Hulskamp, M., K. Schneitz, and R. E. Pruitt. 1995. Genetic evidence for a long-range activity that directs pollen tube guidance in Arabidopsis. Plant Cell **7**:57–64.

Huntzinger, M., R. Karban, T. P. Young, and T. M. Palmer. 2004. Relaxation of induced indirect defenses of acacias following exclusion of mammalian herbivores. Ecology **85**:609–614.

Huq, E. 2006. Degradation of negative regulators: a common theme in hormone and light signaling networks? Trends in Plant Science **11**:4–7.

Hurlbert, S. 1984. Pseudoreplication and the design of ecological field experiments. Ecological Monographs **54**:187–211.

Hutchings, M. J., and H. de Kroon. 1994. Foraging in plants—the role of morphological plasticity in resource acquisition. Advances in Ecological Research **25**:159–238.

Ichimura, K., K. Shinozaki, G. Tena, J. Sheen, Y. Henry, A. Champion, M. Kreis, S. Zhang, H. Hirt, C. Wilson, E. Heberle-Bors, B. E. Ellis, P. C. Morris, R. W. Innes, J. R. Ecker, D. Scheel, D. F. Klessig, Y. Machida, J. Mundy, Y. Ohashi, and J. C. Walker.

2002. Mitogen-activated protein kinase cascades in plants: a new nomenclature. Trends in Plant Science **7**:301–308.

Iqbal, M., and M. Ashraf. 2007. Seed preconditioning modulates growth, ionic relations, and photosynthetic capacity in adult plants of hexaploid wheat under stress. Journal of Plant Nutrition **30**:381–396.

Irwin, R. E., L. S. Adler, and A. K. Brody. 2004. The dual role of floral traits: pollinator attraction and plant defense. Ecology **85**:1503–1511.

Irwin, R. E., J. L. Bronstein, J. S. Manson, and L. Richardson. 2010. Nectar robbing: ecological and evolutionary perspectives. Annual Review of Ecology, Evolution, and Systematics **41**:271–292.

Irwin, R. E., and S. Y. Strauss. 2005. Flower color microevolution in wild radish: evolutionary response to pollinator-mediated selection. American Naturalist **165**:225–237.

Irwin, R. E., S. Y. Strauss, S. Storz, A. Emerson, and G. Guibert. 2003. The role of herbivores in the maintenance of a flower color polymorphism in wild radish. Ecology **84**:1733–1743.

Ito, K., and S. Sakai. 2009. Optimal defense strategy against herbivory in plants: conditions selecting for induced defense, constitutive defense, and no defense. Journal of Theoretical Biology **260**:453–459.

Izaguirre, M. M., C. A. Mazza, M. Biondini, I. T. Baldwin, and C. L. Ballare. 2006. Remote sensing of future competitors: impacts on plant defenses. Proceedings of the National Academy of Sciences **103**:7170–7174.

Izhaki, I., and U. N. Safriel. 1989. Why are there so few exclusively frugivorous birds—experiments on fruit digestibility. Oikos **54**:23–32.

Jablonka, E., and G. Raz. 2009. Transgenerational epigenetic inheritance: prevalence, mechanisms, and implications for the study of heredity and evolution. Quarterly Review of Biology **84**:131–176.

Jackson, R. B., J. H. Manwaring, and M. M. Caldwell. 1990. Rapid physiological adjustment of roots to localized soil enrichment. Nature **344**:58–60.

Jacobs, G. H., J. Neitz, and J. F. Deegan. 1991. Retinal receptors in rodents maximally sensitive to ultraviolet light. Nature **353**:655–656.

Jaffe, M. J., C. Gibson, and R. Biro. 1977. Physiological studies of mechanically stimulated motor responses of flower parts. 1. Characterization of thigmotrophic stamens of *Portulaca grandiflora* Hook. Botanical Gazette **138**:438–447.

Jaffe, M. J., A. C. Leopold, and R. C. Staples. 2002. Thigmo responses in plants and fungi. American Journal of Botany **89**:375–382.

Jaffe, M. J., and M. Shotwell. 1980. Physiological studies on pea tendrils. 11. Storage of tactile sensory information prior to the blue-light activation effect. Physiologia Plantarum **50**:78–82.

Jakab, G., J. Ton, V. Flors, L. Zimmerli, J.-P. Metraux, and B. Mauch-Mani. 2005. Enhancing Arabidopsis salt and drought stress tolerance by chemical priming for its abscisic acid responses. Plant Physiology **139**:267–274.

James, D. G., and T. S. Price. 2004. Field-testing of methyl salicylate for recruitment and retention of beneficial insects in grapes and hops. Journal of Chemical Ecology **30**:1613–1628.

Jansen, R. M. C., M. Miebach, E. Kleist, E. J. van Henten, and J. Wildt. 2009. Release of lipoxygenase products and monoterpenes by tomato plants as an indicator of *Botrytis cinerea*-induced stress. Plant Biology **11**:859–868.

Janson, C. H. 1983. Adaptation of fruit morphology to dispersal agents in a neotropical forest. Science **219**:187–189.

Janzen, D. H. 1966. Coevolution of mutualism between ants and acacias in Central America. Evolution **20**:249–275.

Janzen, D. H. 1967. The interaction of the bull's-horn acacia (*Acacia cornigera* L.) with one of its ant inhabitants (*Pseudomyrmex jerruginea* F. Smith) in eastern Mexico. University of Kansas Science Bulletin **47**:315–558.

Janzen, D. H. 1970. Herbivores and the number of tree species in tropical forests. American Naturalist **104**:501–528.

Janzen, D. H. 1977. Why fruits rot, seeds mold, and meat spoils. American Naturalist **111**:691–713.

Janzen, D. H. 1981. *Enterolobium cyclocarpum* seed passage rate and survival in horses, Costa Rican Pleistocene seed dispersal agents. Ecology **62**:593–601.

Jefferis, G. S. X. E. 2005. Insect olfaction: a map of smell in the brain. Current Biology **15**:R668–R670.

Jersakova, J., and S. D. Johnson. 2006. Lack of floral nectar reduces self-pollination in a fly-pollinated orchid. Oecologia **147**:60–68.

Johnson, M. T. J., A. A. Agrawal, J. Maron, and J.-P. Salminen. 2009. Heritability, covariation and natural selection on 24 traits of common evening primrose (*Oenothera biennis*) from a field experiment. Journal of Evolutionary Biology **22**:1295–1307.

Johnson, R., J. Narvaez, G. H. An, and C. Ryan. 1989. Expression of proteinase inhibitors I and II in transgenic tobacco plants: effects on natural defense against *Manduca sexta* larvae. Proceedings of the National Academy of Sciences **86**:9871–9875.

Johnson, S. D. 2000. Batesian mimicry in the non-rewarding orchid *Disa pulchra*, and its consequences for pollinator behaviour. Biological Journal of the Linnean Society **71**:119–132.

Johnson, S. D. 2006. Pollinator-driven speciation in plants. Pages 295–310 *in* L. D. Harder and S. C. H. Barrett, editors. Ecology and Evolution of Flowers. Oxford University Press, Oxford.

Johnson, S. D., and L. A. Nilsson. 1999. Pollen carryover, geitonogamy, and the evolution of deceptive pollination systems in orchids. Ecology **80**:2607–2619.

Jones, C. G., R. F. Hopper, J. S. Coleman, and V. A. Krischik. 1993. Control of systemically induced herbivore resistance by plant vascular architecture. Oecologia **93**:452–456.

Jones, J. D. G., and J. L. Dangl. 2006. The plant immune system. Nature **444**:323–329.

Jung, H. W., T. J. Tschaplinski, L. Wang, J. Glazebrook, and J. T. Greenberg. 2009. Priming in systemic plant immunity. Science **324**:89–91.

Junker, R. R., and D. Tholl. 2013. Volatile organic compound mediated interactions at the plant-microbe interface. Journal of Chemical Ecology **39**:810–825.

Jurgens, A. 2006. Comparative floral morphometrics in day-flowering, night-flowering and self-pollinated Caryophylloideae (Agrostemma, Dianthus, Saponaria, Silene, and Vaccaria). Plant Systematics and Evolution **257**:233–250.

Kaltz, O., and J. A. Shykoff. 2001. Male and female *Silene latifolia* plants differ in per-contact risk of infection by a sexually transmitted disease. Journal of Ecology **89**:99–109.

Kang, S., H. B. Kim, H. Lee, S. H. Choi, C. J. Oh, S. I. Kwon, and C. S. An. 2006. Overexpression in Arabidopsis of a plasma membrane–targeting glutamate receptor from small radish increases glutamate-mediated Ca^{2+} influx and delays fungal infection. Molecules and Cells **21**:418–427.

Kant, M. R., P. M. Bleeker, M. Van Wijk, R. C. Schuurink, and M. A. Haring. 2009. Plant volatiles in defence. Advances in Botanical Research **51**:613–666.

Kaplan, I. 2012. Attracting carnivorous arthropods with plant volatiles: the future of biocontrol or playing with fire? Biological Control **60**:77–89.

Kappers, I. F., A. Aharoni, T. W. J. M. van Herpen, L. L. P. Luckerhoff, M. Dicke, and H. J. Bouwmeester. 2005. Genetic engineering of terpenoid metabolism attracts, body-guards to Arabidopsis. Science **309**:2070–2072.

Karban, R. 2008. Plant behaviour and communication. Ecology Letters **11**:727–739.

Karban, R. 2011. The ecology and evolution of induced resistance against herbivores. Functional Ecology **25**:339–347.

Karban, R., and F. R. Adler. 1996. Induced resistance to herbivores and the information content of early season attack. Oecologia **107**:379–385.

Karban, R., A. A. Agrawal, and M. Mangel. 1997. The benefits of induced defenses against herbivores. Ecology **78**:1351–1355.

Karban, R., A. A. Agrawal, J. S. Thaler, and L. S. Adler. 1999. Induced plant responses and information content about risk of herbivory. Trends in Ecology and Evolution **14**:443–447.

Karban, R., and I. T. Baldwin. 1997. Induced Responses to Herbivory. University of Chicago Press, Chicago.

Karban, R., I. T. Baldwin, K. J. Baxter, G. Laue, and G. W. Felton. 2000. Communication between plants: induced resistance in wild tobacco plants following clipping of neighboring sagebrush. Oecologia **125**:66–71.

Karban, R., and J. R. Carey. 1984. Induced resistance of cotton seedlings to mites. Science **225**:53–54.

Karban, R., G. English-Loeb, and D. Hougen-Eitzman. 1997. Mite vaccinations for sustainable management of spider mites in vineyards. Ecological Applications **7**:183–193.

Karban, R., M. Huntzinger, and A. C. McCall. 2004. The specificity of eavesdropping on sagebrush by other plants. Ecology **65**:1846–1852.

Karban, R., and J. Maron. 2002. The fitness consequences of interspecific eavesdropping between plants. Ecology **83**:1209–1213.

Karban, R., and J. H. Myers. 1989. Induced plant responses to herbivory. Annual Review of Ecology and Systematics **20**:331–348.

Karban, R., and K. Nagasaka. 2004. Are defenses of wild radish populations well matched with variability and predictability of herbivory? Evolutionary Ecology **18**:283–301.

Karban, R., and C. Niiho. 1995. Induced resistance and susceptibility to herbivory: plant memory and altered development. Ecology **76**:1220–1225.

Karban, R., and K. Shiojiri. 2009. Self-recognition affects plant communication and defense. Ecology Letters **12**:502–506.

Karban, R., K. Shiojiri, M. Huntzinger, and A. C. McCall. 2006. Damage-induced resistance in sagebrush: volatiles are key to intra- and interplant communication. Ecology **87**:922–930.

Karban, R., K. Shiojiri, S. Ishizaki, W. C. Wetzel, and R. Y. Evans. 2013. Kin recognition affects plant communication and defence. Proceedings of the Royal Society B **280**:20123062.

Karban, R., L. H. Yang, and K. F. Edwards. 2014. Volatile communication between plants that affects herbivory: a meta-analysis. Ecology Letters **17**:44-52.

Kathiria, P., C. Sidler, A. Golubov, M. Kalischuk, L. M. Kawchuk, and I. Kovalchuk. 2010. Tobacco mosaic virus infection results in an increase in recombination frequency and resistance to viral, bacterial, and fungal pathogens in the progeny of infected tobacco plants. Plant Physiology **153**:1859–1870.

Kebrom, T. H., and T. P. Brutnell. 2007. The molecular analysis of the shade avoidance syndrome in the grasses has begun. Journal of Experimental Botany **58**:3079–3089.

Keeling, P. J. 2004. Diversity and evolutionary history of plastids and their hosts. American Journal of Botany **91**:1481–1493.

Kegge, W., and R. Pierik. 2010. Biogenic volatile organic compounds and plant competition. Trends in Plant Science **15**:126–132.

Kerwin, R. E., J. M. Jimenez-Gomez, D. Fulop, S. L. Harmer, J. N. Maloof, and D. J. Kliebenstein. 2011. Network quantitative trait loci mapping of circadian clock outputs identifies metabolic pathway-to-clock linkages in Arabidopsis. Plant Cell **23**:471–485.

Kessler, A., and I. T. Baldwin. 2001. Defensive function of herbivore-induced plant volatile emissions in nature. Science:2141–2144.

Kessler, A., and R. Halitschke. 2009. Testing the potential for conflicting selection on floral chemical traits by pollinators and herbivores: predictions and case study. Functional Ecology **23**:901–912.

Kessler, A., R. Halitschke, and I. T. Baldwin. 2004. Silencing the jasmonate cascade: induced plant defenses and insect populations. Science **305**:665–668.

Kessler, A., R. Halitschke, C. Diezel, and I. T. Baldwin. 2006. Priming of plant defense responses in nature by airborne signaling between *Artemisia tridentata* and *Nicotiana attenuata*. Oecologia **148**:280–292.

Kessler, A., and M. Heil. 2011. The multiple faces of indirect defenses and their agents of natural selection. Functional Ecology **25**:348–357.

Kessler, D., K. Gase, and I. T. Baldwin. 2008. Field experiments with transformed plants reveal the sense of floral scents. Science **321**:1200–1202.

Kessmann, H., T. Staub, C. Hofmann, T. Maetzke, J. Herzog, E. Ward, S. Uknes, and J. Ryals. 1994. Induction of systemic acquired disease resistance in plants by chemicals. Annual Review of Phytopathology **32**:439–459.

Keuskamp, D. H., S. Pollmann, L. A. C. J. Voesenek, A. J. M. Peetersa, and R. Pierik. 2010. Auxin transport through PIN-FORMED 3 (PIN3) controls shade avoidance and fitness during competition. Proceedings of the National Academy of Sciences **107**:22740–22744.

Kevan, P. G. 1975. Sun-tracking solar furnaces in high arctic flowers—significance for pollination and insects. Science **189**:723–726.

Khan, Z. R., K. Ampong-Nyarko, P. Chiliswa, A. Hassanaki, S. Kimani, W. Lwande, W. A. Overholt, J. A. Pickett, L. E. Smart, L. J. Wadhams, and C. Woodcock. 1997. Intercropping increases parasitism of pests. Nature **388**:631–632.

Khan, Z. R., D. G. James, C. A. O. Midega, and J. Pickett. 2008. Chemical ecology and conservation biological control. Biological Control **45**:210–214.

Khan, Z. R., C. A. O. Midega, T. J. A. Bruce, A. M. Hooper, and J. A. Pickett. 2010. Exploiting phytochemicals for developing a "push-pull" crop protection strategy for cereal farmers in Africa. Journal of Experimental Botany **61**:4185–4196.

Kiefer, I. W., and A. J. Slusarenko. 2003. The pattern of systemic acquired resistance induction within the Arabidopsis rosette in relation to the pattern of translocation. Plant Physiology **132**:840–847.

Kiers, E. T., M. Duhamel, Y. Beesetty, J. A. Mensah, O. Franken, E. Verbruggen, C. R. Fellbaum, G. A. Kowalchuk, M. M. Hart, A. Bago, T. M. Palmer, S. A. West, P. Vanden-koornhuyse, J. Jansa, and H. Bucking. 2011. Reciprocal rewards stabilize cooperation in the mycorrhizal symbiosis. Science **333**:880–882.

Kiers, E. T., R. A. Rousseau, S. A. West, and R. F. Denison. 2003. Host sanctions and the legume-rhizobium mutualism. Nature **425**:78–81.

Kikuta, S. B., M. A. LoGullo, A. Nardini, H. Richter, and S. Salleo. 1997. Ultrasound acoustic emissions from dehydrating leaves of deciduous and evergreen trees. Plant, Cell and Environment **20**:1381–1390.

Kim, J., H. Quaghebeur, and G. W. Felton. 2011. Reiterative and interruptive signaling in induced plant resistance to chewing insects. Phytochemistry **72**:1624–1634.

Kinoshita, T., M. Doi, S. N., T. Kagawa, M. Wada, and K. Shimazaki. 2001. Phot1 and phot2 mediate blue light regulation of stomatal opening. Nature **414**:656–660.

Klessig, D. F., J. Durner, R. Noad, D. A. Navarre, D. Wendehenne, D. Kumar, J. M. Zhou, J. Shah, S. Zhang, P. Kachroo, Y. Trifa, D. Pontier, E. Lam, and H. Silva. 2000. Nitric oxide and salicylic acid signaling in plant defense. Proceedings of the National Academy of Sciences **97**:8849–8855.

Klooster, M. R., D. L. Clark, and T. M. Culley. 2009. Cryptic bracts facilitate herbivore avoidance in the mycoheterotrophic plant *Monotropsis odorata* (Ericaceae). American Journal of Botany **96**:2197–2205.

Kloppholz, S., H. Kuhn, and N. Requena. 2011. A secreted fungal effector of Glomus intraradices promotes symbiotic biotrophy. Current Biology **21**:1204–1209.

Knight, H., S. Brandt, and M. R. Knight. 1998. A history of stress alters drought calcium signaling pathways in Arabidopsis. Plant Journal **161**:681–687.

Knight, H., A. Trewavas, and M. R. Knight. 1997. Calcium signaling in *Arabidopsis thaliana* responding to drought and salinity. Plant Journal **12**:1067–1078.

Knight, T. A. 1806. On the direction of the radicle and germen during the vegetation of seeds. Philosophical Transactions of the Royal Society of London **99**:108–120.

Knight, T. M., J. M. Chase, H. Hillebrand, and R. D. Holt. 2006. Predation on mutualists can reduce the strength of trophic cascades. Ecology Letters **9**:1173–1178.

Knight, T. M., J. A. Steets, J. C. Vamosi, S. J. Mazer, M. Burd, D. R. Campbell, M. R. Dudash, M. O. Johnston, R. J. Mitchell, and T. L. Ashman. 2005. Pollen limitation of plant reproduction: pattern and process. Annual Review of Ecology and Systematics **36**:467–497.

Knoester, M., L. C. van Loon, J. van den Heuvel, J. Hennig, J. F. Bol, and H. J. M. Linthorst. 1998. Ethylene-insensitive tobacco lacks nonhost resistance against soil-borne fungi. Proceedings of the National Academy of Sciences **95**:1933–1937.

Knudsen, J. T., R. Eriksson, J. Gershenzon, and B. Stahl. 2006. Diversity and distribution of floral scent. Botanical Review **72**:1–120.

Knutson, R. M. 1974. Heat production and temperature regulation in eastern skunk cabbage. Science **186**:746–747.

Kobayashi, A., A. Takahashi, Y. Kakimoto, Y. Miyazawa, N. Fujii, A. Higashitani, and H. Takahashi. 2007. A gene essential for hydrotropism in roots. Proceedings of the National Academy of Sciences **104**:4724–4729.

Kobilinsky, A., A. I. Nazer, and F. Dubois-Brissonnet. 2007. Modeling the inhibition of *Salmonella typhimurium* growth by combination of food antimicrobials. International Journal of Food Microbiology **115**:95–109.

Koda, Y., Y. Kikuta, T. Kithara, T. Nishi, and K. Mori. 1992. Comparison of various biological activities of stereoisomers of methyl jasmonate. Phytochemistry **31**:1111–1114.

Koptur, S. 1992. Extrafloral nectary-mediated interactions between insects and plants. Pages 81–129 *in* E. A. Bernays, editor. Insect-Plant Interactions, vol. 4. CRC Press, Boca Raton, FL.

Korbecka, G., P. G. L. Klinkhamer, and K. Vrieling. 2002. Selective embryo abortion hypothesis revisited—a molecular approach. Plant Biology 4:298–310.

Kozuka, T., G. Horiguchi, G.-T. Kim, M. Ohgishi, T. Sakai, and H. Tsukaya. 2005. The different growth responses of the *Arabidopsis thaliana* leaf blade and the petiole during shade avoidance are regulated by photoreceptors and sugar. Plant and Cell Physiology 46:213–223.

Kramer, M., R. Sanders, H. Bolkan, C. Waters, R. E. Sheeny, and W. R. Hiatt. 1992. Postharvest evaluation of transgenic tomatoes with reduced levels of polygalacturonase: processing, firmness and disease resistance. Postharvest Biology and Technology 1:241–255.

Krouk, G., B. Lacombe, A. Bielach, F. Perrine-Walker, K. Malinska, E. Mounier, K. Hoyerova, P. Tillard, S. Leon, K. Ljung, E. Zazimalova, E. Benkova, P. Nacry, and A. Gojon. 2010. Nitrate-regulated auxin transport by NRT1.1 defines a mechanism for nutrient sensing in plants. Developmental Cell 18:927–937.

Kuc, J. 1987. Plant immunization and its applicability for disease control. Pages 255–274 *in* I. Chet, editor. Innovative Approaches to Plant Disease Control. John Wiley, New York.

Kuc, J. 1995. Induced systemic resistance—an overview. Pages 169–175 *in* R. Hammerschmidt and J. Kuc, editors. Induced Resistance to Disease in Plants. Kluwer, Dordrecht.

Kurepin, L. V., L. J. Walton, D. M. Reid, and C. C. Chinnappa. 2010. Light regulation of endogenous salicylic acid levels in hypocotyls of *Helianthus annuus* seedlings. Botany 88:668–674.

Kurishige, N. S., and A. A. Agrawal. 2005. Phenotypic plasticity to light competition and herbivory in *Chenopodium album* (Chenopodiaceae). American Journal of Botany 92:21–26.

Kwaaitaal, M., R. Huisman, J. Maintz, A. Reinstadler, and R. Panstruga. 2011. Ionotropic glutamate receptor (iGluR)-like channels mediate MAMP-induced calcium influx in *Arabidopsis thaliana*. Biochemical Journal 440:355–365.

Lamb, C., and R. A. Dixon. 1997. The oxidative burst in plant disease resistance. Annual Review of Plant Physiology and Plant Molecular Biology 48:251–275.

Lange, B. M., S. S. Mahmoud, M. R. Wildung, G. W. Turner, E. M. Davis, I. Lange, R. C. Baker, R. A. Boydston, and R. Croteau. 2011. Improving peppermint essential oil yield and composition by metabolic engineering. Proceedings of the National Academy of Sciences 41:16944–16949.

Lankau, R. 2007. Specialist and generalist herbivores exert opposing selection on a chemical defense. New Phytologist 175:176–185.

Lankinen, A., and S. Kiboi. 2007. Pollen donor identity affects timing of stigma receptivity in *Collinsia heterophylla* (Plantaginaceae): a sexual conflict during pollen competition? American Naturalist 170:854–863.

Laothawornkitkul, J., R. M. C. Jansen, H. M. Smid, H. J. Bouwmeester, J. Muller, and A. H. C. van Bruggen. 2010. Volatile organic compounds as a diagnostic marker of late blight infected potato plants: a pilot study. Crop Protection 29:872–878.

Lee, B. H., D. A. Henderson, and J.-K. Zhu. 2005a. The Arabidopsis cold-responsive transcriptome and its regulation by ICE1. Plant Cell 17:3155–3175.

Lee, D., D. H. Polisensky, and J. Braam. 2005b. Genome wide identification of touch and darkness-regulated Arabidopsis genes: a focus on calmodulin-like and XTH genes. New Phytologist **165**:429–444.

Lee, T. D. 1984. Patterns of fruit maturation—a gametophyte competition hypothesis. American Naturalist **123**:427–432.

Lehrer, M., G. A. Horridge, S. W. Zhang, and R. Gadagkar. 1995. Shape vision in bees—innate preference for flower-like patterns. Philosophical Transactions of the Royal Society of London B **347**:123–137.

Lehtila, K. 1996. Optimal distribution of herbivory and localized compensatory responses within a plant. Vegetatio **127**:99–109.

Lemaux, P. G. 2008. Genetically engineered plants and foods: a scientist's analysis of the issues (part I). Annual Review of Plant Biology **59**:771–812.

Lennartsson, T., P. Nilsson, and J. Tuomi. 1998. Induction of overcompensation in the field gentian, *Gentianella campestris*. Ecology **79**:1061–1072.

Leonard, A. S., J. Brent, D. R. Papaj, and A. Dornhaus. 2013. Floral nectar guide patterns discourage nectar robbing by bumble bees. PLoS One **8**:e55914.

Leonard, A. S., A. Dornhaus, and D. R. Papaj. 2011. Forget-me-not: complex floral displays, inter-signal interactions, and pollinator cognition. Current Zoology **57**:215–224.

Leonard, A. S., and D. R. Papaj. 2011. "X" marks the spot: the possible benefits of nectar guides to bees and plants. Functional Ecology **25**:1293–1301.

Lev-Yadun, S. 2009. Aposematic (warning) coloration in plants. Pages 167–202 *in* F. Baluska, editor. Plant-Environment Interactions: From Sensory Plant Biology to Active Plant Behavior. Springer, Berlin.

Lev-Yadun, S., and M. Halpern. 2007. Ergot (*Claviceps purpurea*)—an aposematic fungus. Symbiosis **43**:105–108.

Levey, D. J. 1987. Seed size and fruit-handling techniques of avian frugivores. American Naturalist **129**:471–485.

Levey, D. J., and M. M. Byrne. 1993. Complex ant plant interactions: rain-forest ants as secondary dispersers and postdispersal seed predators. Ecology **74**:1802–1812.

Levitis, D. A., W. Z. Lidicker, and G. Freund. 2009. Behavioural biologists do not agree on what constitutes behaviour. Animal Behaviour **78**:103–110.

Lewinsohn, E., F. Schalechet, J. Wilkinson, K. Matsui, Y. Tadmor, K.-H. Nam, O. Amar, E. Lastochkin, O. Larkov, U. Ravid, W. Hiatt, S. Gepstein, and E. Pichersky. 2001. Enhanced levels of the aroma and flavor compound S-linalool by metabolic engineering of the terpenoid pathway in tomato fruits. Plant Physiology **127**:1256–1265.

Lewis, W. J., and J. H. Tumlinson. 1988. Host detection by chemically mediated associative learning in a parasitic wasp. Nature **331**:257–259.

Li, J., H. Yang, W. A. Peer, G. Richter, J. Blakeslee, A. Bandyopadhyay, B. Titapiwantakun, S. Undurraga, M. Khodakovskaya, E. L. Richards, B. Krizek, A. S. Murphy, S. Gilroy, and R. Gaxiola. 2005. Arabidopsis H+-PPase AVP1 regulates auxin-mediated organ development. Science **310**:121–125.

Li, L., C. Li, G. I. Lee, and G. A. Howe. 2002. Distinct roles for jasmonate synthesis and action in the systemic wound response of tomato. Proceedings of the National Academy of Sciences **99**:6416–6421.

Liang, Y., Y. Cao, K. Tanaka, S. Thibivilliers, J. Wan, J. Choi, C. H. Kang, J. Qiu, and G. Stacey. 2013. Nonlegumes respond to rhizobial Nod factors by suppressing the innate immune response. Science **341**:1384–1387.

Liao, H., H. Wan, J. Shaff, X. Wang, X. Yan, and L. V. Kochian. 2006. Phosphorus and aluminum interactions in soybean in relation to aluminum tolerance, exudation of specific organic acids from different regions of the intact root system. Plant Physiology **141**:674–684.

Lin, Z., S. Zheng, and D. Grierson. 2009. Recent advances in ethylene research. Journal of Experimental Botany **60**:3311–3336.

Lindemann, B. 2001. Receptors and transduction in taste. Nature **413**:219–225.

Linkosalo, T., and M. J. Lechowicz. 2006. Twilight far-red treatment advances leaf bud burst of silver birch (*Betula pendula*). Tree Physiology **26**:1249–1256.

Little, D., C. Gouhier-Darimont, F. Bruessow, and P. Reymond. 2007. Oviposition by pierid butterflies triggers defense responses in Arabidopsis. Plant Physiology **143**:784–800.

Liu, J. Y., L. A. Blaylock, G. Endre, J. Cho, C. D. Town, K. A. VandenBosch, and M. J. Harrison. 2003. Transcript profiling coupled with spatial expression analyses reveals genes involved in distinct developmental stages of an arbuscular mycorrhizal symbiosis. Plant Cell **15**:2106–2123.

Lomascolo, S. B., D. J. Levey, R. T. Kimball, B. M. Bolker, and H. T. Alborn. 2010. Dispersers shape fruit diversity in *Ficus* (Moraceae). Proceedings of the National Academy of Sciences **107**:14668–14672.

Lorrain, S., T. Allen, P. Duek, G. C. Whitelam, and C. Fankhauser. 2008. Phytochrome-mediated inhibition of shade avoidance involves degradation of growth-promoting bHLH transcription factors. Plant Journal **53**:312–323.

Louda, S. M., and J. E. Rodman. 1983a. Concentration of glucosinolates in relation to habitat and insect herbivory for the native crucifer *Cardamine cordifolia*. Biochemical Systematics and Ecology **11**:199–207.

Louda, S. M., and J. E. Rodman. 1983b. Ecological patterns in the glucosinolate content of a native mustard, *Cardamine cordifolia*, in the Rocky Mountains. Journal of Chemical Ecology **9**:397–422.

Louda, S. M., and J. E. Rodman. 1996. Insect herbivory as a major factor in the shade distribution of a native crucifer (*Cardamine cordifolia* A. Gray), bittercress. Journal of Ecology **84**:229–237.

Loughrin, J. H., A. Manukian, R. R. Heath, T. C. J. Turlings, and J. H. Tumlinson. 1994. Diurnal cycle of emission of induced volatile terpenoids by herbivore-injured cotton plants. Proceedings of the National Academy of Sciences **91**:11836–11840.

Lowenberg, G. J. 1994. Effects of floral herbivory on maternal reproduction in *Sanicula arctopoides* (Apiaceae). Ecology **75**:359–369.

Lowman, M. D. 1982. Effects of different rates and methods of leaf area removal on rain forest seedlings of coachwood (*Ceratopetalulm apetalum*). Australian Journal of Botany **30**:477–483.

Luna, E., T. J. A. Bruce, M. R. Roberts, V. Flors, and J. Ton. 2012. Next-generation systemic acquired resistance. Plant Physiology **158**:844–853.

Lunau, K. 1992. A new interpretation of flower guide colouration: absorption of ultraviolet light enhances colour saturation. Plant Systematics and Evolution **183**:51–65.

Lunau, K. 2000. The ecology and evolution of visual pollen signals. Plant Systematics and Evolution **222**:89–111.

Lunau, K., S. Wacht, and L. Chittka. 1996. Colour choices of naive bees and their implications for colour perception. Journal of Comparative Physiology A **178**:477–489.

Lundberg, D. S., S. L. Lebeis, S. H. Paredes, S. Yourstone, J. Gehring, S. Malfatti, J. Trembleay, A. Engelbrektson, V. Kunin, T. G. del Rio, R. C. Edgar, T. Eickhorst, R. E. Ley, P. Hugenholtz, S. G. Tringe, and J. L. Dangl. 2012. Defining the core *Arabidopsis thaliana* root microbiome. Nature **488**:86–90.

Lundstrom, A. N. 1887. Pflanzenbiologische Studien. 11. Die Anpassungen der Pflanzen an Thiere. Nova Acta Regiae Societatis Scientiarum Upsaliensis ser. 3 **13**:1–87.

Lyon, G. 2007. Agents that can elicit induced resistance. Pages 9–29 in D. Walters, A. Newton, and G. Lyon, editors. Induced Resistance for Plant Defense: A Sustainable Approach to Crop Protection. Blackwell, Oxford.

Macpherson, L. J., S. W. Hwang, T. Miyamoto, A. E. Dubin, A. Patapoutiana, and G. M. Story. 2006. More than cool: promiscuous relationships of menthol and other sensory compounds. Molecular and Cellular Neuroscience **32**:335–343.

Maddox, G. D., and R. B. Root. 1987. Resistance to 16 diverse species of herbivorous insects within a population of goldenrod, *Solidago altissima*: genetic variability and heritability. Oecologia **72**:8–14.

Madjidian, J. A., S. Hydbom, and A. Lankinen. 2012. Influence of number of pollinations and pollen load size on maternal fitness costs in *Collinsia heterophylla*: implications for existence of a sexual conflict over timing of stigma receptivity. Journal of Evolutionary Biology **25**:1623–1635.

Maes, L., F. C. W. Van Nieuwerburgh, Y. Zhang, D. W. Reed, J. Pollier, S. R. F. Vande Casteele, D. Inze, P. S. Covello, D. L. D. Deforce, and A. Goossens. 2011. Dissection of the phytohormonal regulation of trichome formation and biosynthesis of the antimalarial compound artemisinin in *Artemisia annua* plants. New Phytologist **189**:175–189.

Maffei, M. E., S. Bossi, D. Spiteller, A. Mithofer, and W. Boland. 2004. Effects of feeding *Spodoptera littoralis* on lima bean leaves. I. Membrane potentials, intracellular calcium variations, oral secretions, and regurgitate components. Plant Physiology **134**:1752–1762.

Maffei, M. E., J. Gertsch, and G. Appendino. 2011. Plant volatiles: production, function and pharmacology. Natural Product Reports **28**:1359–1380.

Maffei, M. E., A. Mithofer, and W. Boland. 2007. Before gene expression: early events in plant-insect interaction. Trends in Plant Science **12**:310–316.

Mahall, B. E., and R. M. Callaway. 1991. Root communication among desert shrubs. Proceedings of the National Academy of Sciences **88**:874–876.

Maischak, H., P. A. Grigoriev, H. Vogel, W. Boland, and A. Mithofer. 2007. Oral secretions from herbivorous lepidopteran larvae exhibit ion channel–forming activities. FEBS Letters **581**:898–904.

Malamy, J., J. P. Carr, D. F. Klessig, and I. Raskin. 1990. Salicylic acid: a likely endogenous signal in the resistance response of tobacco to viral infection. Science **250**:1002–1004.

Mantyla, E., G. A. Alessio, J. D. Blande, J. Heijari, J. K. Holopainen, T. Laaksonen, P. Piirtola, and T. Klemola. 2008. From plants to birds: higher avian predation rates in trees responding to insect herbivory. PLoS One **3**:e2832.

Marder, M. 2012. Plant intentionality and the phenomenological framework of plant intelligence. Plant Signaling and Behavior **7**:1–8.

Marder, M. 2013. Plant intelligence and attention. Plant Signaling and Behavior **8**:e23902.

Margulis, L. 1971. Origin of plant and animal cells. American Scientist **59**:230–235.

Marquis, R. J. 1992. A bite is a bite is a bite? Constraints on response to folivory in *Piper arieianum* (Piperaceae). Ecology **73**:143–152.

Marshall, D. L. 1988. Postpollination effects on seed paternity: mechanisms in addition to microgametophyte competition operate in wild radish. Evolution **42**:1256–1266.

Marshall, D. L., and M. W. Folsom. 1991. Mate choice in plants: an anatomical to population perspective. Annual Review of Ecology and Systematics **22**:37–63.

Marten-Rodriguez, S., A. Almarales-Castro, and C. B. Fenster. 2009. Evaluation of pollination syndromes in Antillean Gesneriaceae: evidence for bat, hummingbird and generalized flowers. Journal of Ecology **97**:348–359.

Martineau, B. 2001. First Fruit: The Creation of the Flavr Savr Tomato and the Birth of Genetically Engineered Food. McGraw-Hill, New York.

Maschinski, J., and T. G. Whitham. 1989. The continuum of plant responses to herbivory: the influence of plant association, nutrient availability and timing. American Naturalist **134**:1–19.

Matsui, K. 2006. Green leaf volatiles: hydroperoxide lyase pathway of oxylipin metabolism. Current Opinion in Plant Biology **9**:274–280.

Matthes, M. C., T. J. A. Bruce, J. Ton, P. J. Verrier, J. A. Pickett, and J. A. Napier. 2010. The transcriptome of cis-jasmone-induced resistance in *Arabidopsis thaliana* and its role in indirect defence. Planta **232**:1163–1180.

Mattiacci, L., M. Dicke, and M. A. Posthumus. 1995. Beta-glucosidase—an elicitor of herbivore-induced plant odor that attracts host-searching parasitic wasps. Proceedings of the National Academy of Sciences **92**:2036–2040.

Mauricio, R., M. D. Bowers, and F. A. Bazzaz. 1993. Pattern of leaf damage affects fitness of the annual plant *Raphanus sativus* (Brassicaceae). Ecology **74**:2066–2071.

Maynard Smith, J., and D. G. C. Harper. 1995. Animal signals: models and terminology. Journal of Theoretical Biology **177**:305–311.

Maynard Smith, J., and D. G. C. Harper. 2003. Animal Signals. Oxford University Press, Oxford.

McCall, A. C., and R. E. Irwin. 2006. Florivory: the intersection of pollination and herbivory. Ecology Letters **9**:1351–1365.

McCall, A. C., S. J. Murphy, C. Venner, and M. Brown. 2013. Florivores prefer white versus pink petal color morphs in wild radish, *Raphanus sativus*. Oecologia **172**:189–195.

McClure, B., F. Cruz-Garcia, and C. Romero. 2011. Compatibility and incompatibility in S-RNase-based systems. Annals of Botany **108**:647–658.

McConn, M., R. A. Creelman, E. Bell, J. E. Mullet, and J. Browse. 1997. Jasmonate is essential for insect defense in Arabidopsis. Proceedings of the National Academy of Sciences **94**:5473–5477.

McDonald, A. J. S., and W. J. Davies. 1996. Keeping in touch: responses of the whole plant to deficits in water and nitrogen supply. Advances in Botanical Research **22**:229–300.

McKey, D. 1975. The ecology of coevolved seed dispersal systems. Pages 159–191 *in* L. E. Gilbert and P. H. Raven, editors. Coevolution of Animals and Plants. University of Texas Press, Austin, Texas.

Meeuse, B. J. D. 1975. Thermogenic respiration in aroids. Annual Review of Plant Physiology and Plant Molecular Biology **26**:117–126.

Melendez-Ackerman, E., and D. R. Campbell. 1998. Adaptive significance of flower color and inter-trait correlations in an *Ipomopsis* hybrid zone. Evolution **52**:1293–1303.

Menzel, R., and U. Muller. 1996. Learning and memory in honeybees: from behavior to neural substrates. Annual Review of Neuroscience **19**:379–404.

Metlen, K. L., E. T. Aschehoug, and R. M. Callaway. 2009. Plant behavioural ecology: dynamic plasticity in secondary metabolites. Plant, Cell and Environment **32**:641–653.

Metraux, J.-P., H. Signer, J. A. Ryals, E. Ward, M. Wyss-Benz, J. Gaudin, K. Raschdorf, E. Schmid, W. Blum, and B. Inverardi. 1990. Increase in salicylic acid at the onset of systemic acquired resistance in cucumber. Science **250**:1004–1006.

Meyer, K. M., L. L. Soldaat, H. Auge, H.-H. Thulke. 2014. Adaptive and selective seed abortion reveals complex conditional decision making in plants. American Naturalist **183**:376-383.

Michaels, S. D., and R. M. Amasino. 2000. Memories of winter: vernalization and the competence to flower. Plant, Cell and Environment **23**:1145–1153.

Michard, E., P. T. Lima, F. Borges, A. C. Silva, M. T. Portes, J. E. Carvalho, M. Gilliham, L.-H. Liu, G. Obermeyer, and J. A. Feijo. 2011. Glutamate receptor–like genes form Ca^{2+} channels in pollen tubes and are regulated by pistil D-serine. Science **332**:434–437.

Milan, N. F., B. Z. Kacosh, and T. A. Schlenke. 2012. Alcohol consumption as self-medication against blood-borne parasites in the fruit fly. Current Biology **22**:488–493.

Miller, G., K. Schlauch, R. Tam, D. Cortes, M. A. Torres, V. Shulaev, J. L. Dangl, and R. Mittler. 2009. The plant NADPH oxidase RBOHD mediates rapid systemic signaling in response to diverse stimuli. Science Signaling **2**:A26–A35.

Mirabella, R., H. Rauwerda, E. A. Struys, C. Jakobs, C. Triantaphylides, M. A. Haring, and R. C. Schuurink. 2008. The Arabidopsis her1 mutant implicates GABA in E-2-hexenal responsiveness. Plant Journal **53**:197–213.

Mitchell, R. J. 2004. Heritability of nectar traits: why do we know so little? Ecology **85**:1527–1533.

Mitchell, R. J., R. J. Flanagan, B. J. Brown, N. M. Waser, and J. D. Karron. 2009. New frontiers in competition for pollination. Annals of Botany **103**:1403–1413.

Mithofer, A., and W. Boland. 2008. Recognition of herbivory-associated molecular patterns. Plant Physiology **146**:825–831.

Mithofer, A., and W. Boland. 2012. Plant defense against herbivores: chemical aspects. Annual Review of Plant Biology **63**:431–450.

Mithofer, A., G. Wanner, and W. Boland. 2005. Effects of feeding *Spodoptera littoralis* on lima bean leaves. II. Continuous mechanical wounding resembling insect feeding is sufficient to elicit herbivory-related volatile emission. Plant Physiology **137**:1160–1168.

Moehs, C. P., L. Tian, K. W. Osteryoung, and D. DellaPenna. 2001. Analysis of carotenoid biosynthetic gene expression during marigold petal development. Plant Molecular Biology **45**:281–293.

Molinier, J., G. Ries, C. Zipfel, and B. Hohn. 2006. Transgeneration memory of stress in plants. Nature **442**:1046–1049.

Monshausen, G. B., and S. Gilroy. 2009. Feeling green: mechanosensing in plants. Trends in Cell Biology **19**:228–235.

Montalvo, A. M. 1992. Relative success of self and outcross pollen comparing mixed-donor and single-donor pollinations in *Aquilegia caerulea*. Evolution **46**:1181–1198.

Morales, C. L., and A. Traveset. 2008. Interspecific pollen transfer: magnitude, prevalence and consequences for plant fitness. Critical Reviews in Plant Sciences **27**:221–238.

Moran, N. A. 2002. The ubiquitous and varied role of infection in the lives of animals and plants. American Naturalist **160**:S1–S8.

Moran, N. A. 2012. Microbial symbiosis and evolution. Pages 191-196 *in* R. Kolter and S. Maloy, editors. Microbes and Evolution: The World That Darwin Never Saw. ASM Press, Washington, D. C.

Morelli, G., and I. Ruberti. 2002. Light and shade in the photocontrol of Arabidopsis growth. Trends in Plant Science **7**:399–404.

Moreno, J. E., Y. Tao, J. Chory, and C. L. Ballare. 2009. Ecological modulation of plant defense via phytochrome control of jasmonate sensitivity. Proceedings of the National Academy of Sciences **106**:4935–4940.

Morgan, P. W., S. A. Finlayson, K. L. Childs, J. E. Mullet, and W. L. Rooney. 2002. Opportunities to improve adaptability and yield in grasses: lessons from sorghum. Crop Science **42**:1791–1799.

Morita, M. T. 2010. Directional gravity sensing in gravitropism. Annual Review of Plant Biology **61**:705–720.

Morris, W. F. 1996. Mutualism denied? Nectar-robbing bumble bees do not reduce female or male success of bluebells. Ecology **77**:1451–1462.

Morris, W. F., D. P. Vazquez, and N. P. Chacoff. 2010. Benefit and cost curves for typical pollination mutualisms. Ecology **91**:1276–1285.

Mousavi, S. A. R., A. Chauvin, F. Pascaud, S. Kellenberger, and E. E. Farmer. 2013. Glutamate receptor-like genes mediate leaf-to-leaf wound signaling. Nature **500**:422–426.

Muller, R., L. Nilsson, C. Krintel, and T. H. Nielsen. 2004. Gene expression during recovery from phosphate starvation in roots and shoots of *Arabidopsis thaliana*. Physiologia Plantarum **122**:233–243.

Murray, K. G., S. Russell, C. M. Picone, K. Winnett-Murray, W. Sherwood, and M. L. Kuhlmann. 1994. Fruit laxatives and seed passage rates in frugivores: consequences for plant reproductive success. Ecology **75**:989–994.

Mutikainen, P., M. Walls, and J. Ovaska. 1996. Herbivore-induced resistance in *Betula pendula*: the role of plant vascular architecture. Oecologia **108**:723–727.

Myers, J. H., and K. S. Williams. 1984. Does tent caterpillar attack reduce the food quality for red alder foliage? Oecologia **62**:74–79.

Nagy, K. A., D. K. Odell, and R. S. Seymour. 1972. Temperature regulation by inflorescences of *Philodendron*. Science **178**:1195–1197.

Naug, D., and H. S. Arathi. 2007. Receiver bias for exaggerated signals in honeybees and its implications for the evolution of floral displays. Biology Letters **3**:635–637.

Newman, D. J., and G. M. Cragg. 2012. Natural products as sources of new drugs over the 30 years from 1981 to 2010. Journal of Natural Products **75**:311–335.

Nibau, C., D. J. Gibbs, and J. C. Coates. 2008. Branching out in new directions: the control of root architecture by lateral root formation. New Phytologist **179**:595–614.

Nick, P., and E. Schafer. 1988. Spatial memory during the tropism of maize (*Zea mays* L.) coleoptiles. Planta **175**:380–388.

Nicklen, E. F., and D. Wagner. 2006. Conflict resolution in an ant-plant interaction: *Acacia constricta* traits reduce ant costs to reproduction. Oecologia **148**:81–87.

Ninkovic, V. 2003. Volatile communication between barley plants affects biomass allocation. Journal of Experimental Botany **54**:1931–1939.

Nobel, P. S. 2009. Physiological and Environmental Plant Physiology, 4th ed. Academic Press, San Diego.

Noordermeer, M. A., G. A. Veldink, and J. F. G. Bliegenthart. 2001. Fatty acid hydroperoxide lyase: a plant cytochrome P450 enzyme involved in wound healing and pest resistance. Chembiochem **2**:494–504.

Norton, A. P., G. English-Loeb, D. Gadoury, and R. C. Seem. 2000. Mycophagous mites and foliar pathogens: leaf domatia mediate tritrophic interactions in grapes. Ecology **81**:490–499.

Novoplansky, A. 1991. Developmental responses of *Portulaca* seedlings to conflicting spectral signals. Oecologia **88**:138–140.

Novoplansky, A., D. Cohen, and T. Sachs. 1990. How *Portulaca* seedlings avoid their neighbors. Oecologia **82**:490–493.

Nurnberger, T., F. Brunner, K. B., and L. Piater. 2004. Innate immunity in plants and animals: striking similarities and obvious differences. Immunological Reviews **198**:249–266.

Nuttman, C., and P. Willmer. 2003. How does insect visitation trigger floral colour change? Ecological Entomology **28**:467–474.

Nykanen, H., and J. Koricheva. 2004. Damage-induced changes in woody plants and their effects on insect herbivore performance: a meta-analysis. Oikos **104**:247–268.

Nystrand, O., and A. Granstrom. 1997. Post-dispersal predation on *Pinus sylvestris* seeds by *Fringilla* spp.: ground substrate affects selection for seed color. Oecologia **110**:353–359.

O'Connell, L. M., and M. O. Johnston. 1998. Male and female pollination success in a deceptive orchid, a selection study. Ecology **79**:1246–1260.

O'Donnell, P. J., C. Calvert, R. Atzorn, C. Wasternack, H. M. O. Leyser, and D. J. Bowles. 1996. Ethylene as a signal mediating the wound response of tomato plants. Science **274**:1914–1917.

O'Dowd, D. J., and R. W. Pemberton. 1994. Leaf domatia in Korean plants: floristics, frequency, and biogeography. Vegetatio **114**:137–149.

O'Dowd, D. J., and M. F. Willson. 1989. Leaf domatia and mites on Australasian plants—ecological and evolutionary implications. Biological Journal of the Linnean Society **37**:191–236.

Okamoto, M., J. J. Vidmar, and A. D. M. Glass. 2003. Regulation of NRT1 and NRT2 gene families of *Arabidopsis thaliana*: responses to nitrate provision. Plant and Cell Physiology **44**:304–317.

Okuda, S., H. Tsutsui, K. Shiina, et al. 2009. Defensin-like polypeptide LUREs are pollen tube attractants secreted from synergid cells. Nature **458**:357–361.

Oldroyd, G. E. D. 2013. Speak, friend, and enter: signalling systems that promote beneficial symbiotic associations in plants. Nature Reviews Microbiology **11**:252–263.

Oldroyd, G. E. D., J. D. Murray, P. S. Poole, and J. A. Downie. 2011. The rules of engagement in the legume-rhizobial symbiosis. Annual Review of Genetics **45**:119–144.

Ollerton, J., R. Alarcon, N. M. Waser, M. V. Price, S. Watts, L. Cranmer, A. Hingston, C. I. Peter, and J. Rotenberry. 2009. A global test of the pollination syndrome hypothesis. Annals of Botany **103**:1471–1480.

Olsen, J. E., O. Junttila, J. Nilsen, M. E. Eriksson, I. Martinussen, O. Olsson, G. Sandberg, and T. Moritz. 1997. Ectopic expression of oat phytochrome A in hybrid aspen changes critical daylength for growth and prevents cold acclimatization. Plant Journal **12**:1339–1350.

O'Malley, R. C., F. I. Rodriguez, J. J. Esch, B. M. Binder, P. O'Donnell, H. J. Klee, and A. B. Bleecker. 2005. Ethylene-binding activity, gene expression levels, and receptor system output for ethylene receptor family members from Arabidopsis and tomato. Plant Journal **41**:651–659.

Orians, C. 2005. Herbivores, vascular pathways, and systemic induction: facts and artifacts. Journal of Chemical Ecology **31**:2231–2242.

Orians, C. M., M. Ardon, and B. A. Mohammad. 2002. Vascular architecture and patchy nutrient availability generate within-plant heterogeneity in plant traits important to herbivores. American Journal of Botany **89**:270–278.

Orians, C. M., and C. G. Jones. 2001. Plants as mosaics: a functional model for predicting patterns of within-plant resource heterogeneity to consumers based on vascular architecture and local environmental variability. Oikos **94**:493–504.

Orians, C. M., J. Pomerleau, and R. Ricco. 2000. Vascular architecture generates fine scale variation in systemic induction of proteinase inhibitors in tomato. Journal of Chemical Ecology **26**:471–485.

Orrock, J. L. 2013. Exposure of unwounded plants to chemical cues associated with herbivores leads to exposure-dependent changes in subsequent herbivore attack. PLoS One **8**:e79900.

Otte, D. 1974. Effects and functions in the evolution of signaling systems. Annual Review of Ecology and Systematics **4**:385–417.

Paige, K. N. 1992. Overcompensation in response to mammalian herbivory: from mutualistic to antagonistic interactions. Ecology **73**:2076–2085.

Pallini, A., A. Janssen, and M. W. Sabelis. 1997. Odour-mediated responses of phytophagous mites to conspecifics and heterospecific competitors. Oecologia **110**:179–185.

Palmer, T. M., D. F. Doak, M. L. Stanton, J. L. Bronstein, E. T. Kiers, T. P. Young, J. R. Goheen, and R. M. Pringle. 2010. Synergy of multiple partners, including freeloaders, increases host fitness in a multispecies mutualism. Proceedings of the National Academy of Sciences **107**:17234–17239.

Palmer, T. M., M. L. Stanton, T. P. Young, J. R. Goheen, R. M. Pringle, and R. Karban. 2008. Breakdown of an ant-plant mutualism follows the loss of large herbivores from an African savanna. Science **319**:192–195.

Pare, P. W., and J. H. Tumlinson. 1999. Plant volatiles as a defense against insect herbivores. Plant Physiology **121**:325–331.

Park, S.-W., E. Kaimoyo, D. Kumar, S. Mosher, and D. F. Klessig. 2007. Methyl salicylate is a critical mobile signal for plant systemic acquired resistance. Science **318**:113–116.

Pearce, G., D. Strydom, S. Johnson, and C. A. Ryan. 1991. A polypeptide from tomato leaves induces wound-inducible proteinase inhibitor proteins. Science **253**:895–898.

Pearcy, R. W. and D. A. Sims. 1994. Photosynthetic acclimation to changing light environments: scaling from the leaf to the whole plant. Pages 145–174 in M. W. Caldwell and R. W. Pearcy, editors. Exploitation of Environmental Heterogeneity by Plants. Academic Press, San Diego.

Pearcy, R. W., R. L. Chazdon, L. J. Gross, and K. A. Mott. 1994. Photosynthetic utilization of sunflecks: a temporally patchy resource on a time scale of seconds to minutes. Pages 175–208 in M. W. Caldwell and R. W. Pearcy, editors. Exploitation of Environmental Heterogeneity by Plants. Academic Press, San Diego.

Peiffer, M., J. F. Tooker, D. S. Luthe, and G. W. Felton. 2009. Plants on early alert: glandular trichomes as sensors for insect herbivores. New Phytologist **184**:644–656.

Peng, J., J. J. A. van Loon, S. Zheng, and M. Dicke. 2011. Herbivore-induced volatiles in cabbage (*Brassica oleracea*) prime defence responses in neighbouring intact plants. Plant Biology **13**:276–284.

Pereira, H. M., P. W. Leadley, V. Proenca, et al. 2010. Scenarios for global biodiversity in the 21st century. Science **330**:1496–1501.

Peter, C. I., and S. D. Johnson. 2008. Mimics and magnets: the importance of color and ecological facilitation in floral deception. Ecology **89**:1583–1595.

Pettersson, J., J. A. Pickett, B. J. Pye, A. Quiroz, L. E. Smart, L. J. Wadhams, and C. M. Woodcock. 1994. Winter host component reduces colonization by bird-cherry aphid, *Rhopalosiphum padi* (L.) (Homoptera, Aphididae), and other aphids in cereal fields. Journal of Chemical Ecology **20**:2565–2574.

Pfeiffer, M., H. Huttenlocher, and M. Ayasse. 2010. Myrmecochorous plants use chemical mimicry to cheat seed-dispersing ants. Functional Ecology **24**:545–555.

Pichersky, E., J. P. Noel, and N. Dudareva. 2006. Biosynthesis of plant volatiles: nature's diversity and ingenuity. Science **311**:808–811.

Pickett, J. A., G. I. Aradottir, M. A. Birkett, T. J. A. Bruce, A. M. Hooper, C. A. O. Midega, H. D. Jones, M. C. Matthes, J. A. Napier, J. O. Pittchar, L. E. Smart, C. M. Woodcock, and Z. R. Khan. 2014. Delivering sustainable crop protection systems via the seed: exploiting natural constitutive and inducible defence pathways. Philosophical Transactions of the Royal Society B. **369**:20120281.

Pierik, R., M. L. C. Cuppens, L. A. C. J. Voesenek, and E. J. W. Visser. 2004. Interactions between ethylene and gibberellins in phytochrome-mediated shade avoidance responses in tobacco. Plant Physiology **136**:2928–2936.

Pierik, R., L. Mommer, and L. A. C. J. Voesenek. 2013. Molecular mechanisms of plant competition: neighbor detection and response strategies. Functional Ecology **27**:841-853.

Pierik, R., D. Tholena, H. Poorter, E. J. W. Visser, and L. A. C. J. Voesenek. 2006. The Janus face of ethylene: growth inhibition and stimulation. Trends in Plant Science **11**:176–183.

Pierik, R., E. J. W. Visser, H. De Kroon, and L. A. C. J. Voesenek. 2003. Ethylene is required in tobacco to successfully compete with proximate neighbours. Plant, Cell and Environment **26**:1229–1234.

Pierik, R., G. C. Whitelam, L. A. C. J. Voesenek, H. de Kroon, and E. J. W. Visser. 2004. Canopy studies on ethylene-insensitive tobacco identify ethylene as a novel element in blue light and plant-plant signalling. Plant Journal **38**:310–319.

Pieterse, C. M. J., A. Leon-Reyes, S. Van der Ent, and S. C. M. Van Wees. 2009. Networking by small-molecule hormones in plant immunity. Nature Chemical Biology **5**:308–316.

Pinto, D. M., A.-M. Nerg, and J. K. Holopainen. 2007. The role of ozone-reactive compounds, terpenes, and green leaf volatiles (GLVs), in the orientation of *Cotesia plutellae*. Journal of Chemical Ecology **33**:2218–2228.

Plateau, F. 1901. Observations sur le phenomene de la constance chez quelques hymenopteres. Annales de la Societe Entomologique de Belgique **45**:56–83.

Poorter, H., and O. Nagel. 2000. The role of biomass allocation in the growth response of plants to different levels of light, CO_2, nutrients and water: a quantitative review. Australian Journal of Plant Physiology **27**:595–607.

Porter, S. S. 2013. Adaptive divergence in seed color camouflage in contrasting soil environments. New Phytologist **197**:1311–1320.

Poulter, N. S., M. J. Wheeler, M. Bosch, and V. E. Franklin-Tong. 2010. Self-incompatibility in *Papaver*: identification of the pollen S-determinant PrpS. Biochemical Society Transactions **38**:588–592.

Pozo, M. J., and C. Azcon-Aguilar. 2007. Unraveling mycorrhiza-induced resistance. Current Opinion in Plant Biology **10**:393–398.

Preston, C. A., G. Laue, and I. T. Baldwin. 2004. Plant-plant signaling: application of *trans*- or *cis*-methyl jasmonate equivalent to sagebrush releases does not elicit direct defenses in native tobacco. Journal of Chemical Ecology **30**:2193–2214.

Price, P. W. 1988. An overview of organismal interactions in ecosystems in evolutionary and ecological time. Agriculture, Ecosystems and Environment **24**:369-377.

Price, P. W., C. E. Bouton, P. Gross, B. A. McPherson, J. N. Thompson, and A. E. Weis. 1980. Interactions among three trophic levels—influence of plants on interactions between insect herbivores and natural enemies. Annual Review of Ecology and Systematics **11**:41–65.

Provenza, F. D. 1995. Postingestive feedback as an elementary determinant of food preference and intake in ruminants. Journal of Range Management **48**:2–17.

Radhika, V., C. Kost, A. Mithofer, and W. Boland. 2010. Regulation of extrafloral nectar secretion by jasmonates in lima bean is light dependent. Proceedings of the National Academy of Sciences **107**:17228–17233.

Raguso, R. A. 2008. Wake up and smell the roses: the ecology and evolution of floral scent. Annual Review of Ecology and Systematics **39**:549–569.

Raine, N. E., and L. Chittka. 2007. The adaptive significance of sensory bias in a foraging context: floral colour preferences in the bumblebee *Bombus terrestris*. PLoS One **2**:e556.

Raine, N. E., P. Willmer, and G. N. Stone. 2002. Spatial structuring and floral avoidance behavior prevent ant-pollinator conflict in a Mexican ant-acacia. Ecology **83**:3086–3096.

Ramirez, S. R., T. Eltz, M. K. Fujiwara, G. Gerlach, B. Goldman-Huertas, N. D. Tsutsui, and N. E. Pierce. 2011. Asynchronous diversification in a specialized plant-pollinator mutualism. Science **333**:1742–1746.

Raskin, I. 1992. Role of salicylic acid in plants. Annual Review of Plant Physiology and Plant Molecular Biology **43**:439–463.

Rasmann, S., M. De Vos, C. L. Casteel, D. Tian, R. Halitschke, J. Y. Sun, A. A. Agrawal, G. W. Felton, and G. Jander. 2012. Herbivory in the previous generation primes plants for enhanced insect resistance. Plant Physiology **158**:854–863.

Rasmann, S., A. C. Erwin, R. Halitschke, and A. A. Agrawal. 2011. Direct and indirect root defences of milkweed (*Asclepias syriaca*): trophic cascades, trade-offs and novel methods for studying subterranean herbivory. Journal of Ecology **99**:16–25.

Rasmann, S., T. G. Kollner, J. Degenhardt, I. Hiltpold, S. Toepfer, U. Kuhlmann, J. Gershenzon, and T. C. J. Turlings. 2005. Recruitment of entomopathogenic nematodes by insect-damaged maize roots. Nature **434**:732–737.

Rasmussen, J. B., R. Hammerschmidt, and M. N. Zook. 1991. Systemic induction of salicylic acid accumulation in cucumber after inoculation with *Pseudomonas syringue* pv. syringue. Plant Physiology **97**:1342–1347.

Rausher, M. D. 1992. Natural selection and the evolution of plant-insect interactions. Pages 20–88 *in* B. Roitberg and M. B. Isman, editors. Insect Chemical Ecology. Chapman and Hall, New York.

Rausher, M. D. 2008. Evolutionary transitions in floral color. International Journal of Plant Sciences **169**:7-21.

Reddy, A. S. N., G. S. Ali, H. Celesnik, and I. S. Day. 2011. Coping with stresses: roles of calcium- and calcium/calmodulin-regulated gene expression. Plant Cell **23**:2010–2032.

Reid, N. 1991. Coevolution of mistletoes and frugivorous birds? Australian Journal of Ecology **16**:457–469.

Renner, S. S. 2006. Rewardless flowers in the Angiosperms and the role of insect cognition in their evolution. Pages 123–144 *in* N. M. Waser and J. Ollerton, editors. Plant-Pollinator Interactions: From Specialization to Generalization. University of Chicago Press, Chicago.

Retana, J., F. X. Pico, and A. Rodrigo. 2004. Dual role of harvesting ants as seed predators and dispersers of a non-myrmechorous Mediterranean perennial herb. Oikos **105**:377–385.

Reynolds, H. L., and C. D'Antonio. 1996. The ecological significance of plasticity in root weight ratio in response to nitrogen: opinion. Plant and Soil **185**:75–97.

Rhoades, D. F. 1983. Responses of alder and willow to attack by tent caterpillars and webworms: evidence for phenomonal sensitivity of willows. Pages 55–68 *in* P. A. Hedin, editor. Plant Resistance to Insects, Symposium Series 208. American Chemical Society, Washington, D. C.

Rico-Gray, V., and P. S. Oliveira. 2007. The Ecology and Evolution of Ant-Plant Interactions. University of Chicago Press, Chicago.

Rizzini, L., J.-J. Favory, C. Cloix, D. Farrionato, A. O'Hara, E. Kaiserli, R. Baumeiser, E. Schafer, F. Nagy, G. Jenkins, and R. Ulm. 2011. Perception of UV-B by the Arabidopsis UVR8 protein. Science **332**:103–106.

Roberts, D. A. 1983. Acquired resistance to tobacco mosaic virus transmitted to the progeny of hypersensitive tobacco. Virology **124**:161–163.

Robertson, C. 1895. The philosophy of flower seasons, and the phaenological relations of the entomophilous flora and the anthophilous insect fauna. American Naturalist **29**:97–117.

Robinson, D., A. Hodge, B. S. Griffiths, and A. H. Fitter. 1999. Plant root proliferation in nitrogen-rich patches confers competitive advantage. Proceedings of the Royal Society B **266**:431–435.

Rodriguez, A., B. Alquezar, and L. Pena. 2012. Fruit aromas in mature fleshy fruits as signals of readiness for predation and seed dispersal. New Phytologist **197**:36–48.

Rodriguez, A., V. San Andres, M. Cervera, R. A., B. Alquezar, T. Shimada, J. Gadea, M. J. Rodrigo, L. Zacarias, L. Palou, M. M. Lopez, P. Castanera, and L. Pena. 2011. Terpene down-regulation in orange reveals the role of fruit aromas in mediating interactions with insect herbivores and pathogens. Plant Physiology **156**:793–802.

Rodriguez, M. C. S., M. Petersen, and J. Mundy. 2010. Mitogen-activated protein kinases signaling in plants. Annual Review of Plant Biology **61**:621–649.

Rodriguez-Saona, C. R., L. E. Rodriguez-Saona, and C. J. Frost. 2009. Herbivore-induced volatiles in the perennial shrub, *Vaccinium corymbosum*, and their role in inter-branch signaling. Journal of Chemical Ecology **35**:163–175.

Rodriguez-Saona, C., and J. S. Thaler. 2005. The jasmonate pathway alters herbivore feeding behavior: consequences for plant defences. Entomologia Experimentalis et Applicata **115**:125–134.

Roelfsema, M. R. G., and R. Hedrich. 2005. In the light of stomatal opening: new insights into "the Watergate." New Phytologist **167**:665–691.

Roitberg, B., M. L. Reid, and C. Li. 1993. Choosing hosts and mates: the value of learning. Pages 174–194 *in* D. R. Papaj and A. C. Lewis, editors. Insect Learning: Ecology and Evolutionary Perspectives. Chapman and Hall, New York.

Rokhina, E. V., P. Lens, and J. Virkutyte. 2009. Low-frequency ultrasound in biotechnology: state of the art. Trends in Biotechnology **27**:298–306.

Romero, G. A., and C. E. Nelson. 1986. Sexual dimorphism in Catastetem orchids—forcible pollen emplacement and male flower competition. Science **232**:1538–1540.

Romero, G. Q., and W. W. Benson. 2005. Biotic interactions of mites, plants and leaf domatia. Current Opinion in Plant Biology **8**:436–440.

Romero, G. Q., and J. Koricheva. 2011. Contrasting cascade effects of carnivores on plant fitness: a meta-analysis. Journal of Animal Ecology **80**:696–704.

Rosas-Guerrero, V., R. Aguilar, S. Marten-Rodriguez, L. Ashworth, M. Lopezaraiza-Mikel, J. M. Bastida, and M. Quesada. 2014. A quantitative review of pollination syndromes: do floral traits predict effective pollinators? Ecology Letters **17**:388–400.

Rose, U. S. R., A. Manukian, R. R. Heath, and J. H. Tumlinson. 1996. Volatile semio-chemicals released from undamaged cotton leaves—A systemic response of living plants to caterpillar damage. Plant Physiology **111**:487–495.

Ross, A. F. 1961. Systemic acquired resistance induced by localized virus infections in plants. Virology **14**:340–358.

Rossiter, M. C. 1996. Incidence and consequences of inherited environmental effects. Annual Review of Ecology and Systematics **27**:451–476.

Rothschild, M. 1975. Remarks on carotenoids in the evolution of signals. Pages 20–50 *in* L. E. Gilbert and P. H. Raven, editors. Coevolution of Animals and Plants. University of Texas Press, Austin.

Roubik, D. W. 1982. The ecological impact of nectar-robbing bees and pollinating hummingbirds on a tropical shrub. Ecology **63**:354–360.

Ruuhola, T., J.-P. Salminen, S. Haviola, S. Yang, and M. J. Rantala. 2007. Immunological memory of mountain birches: effects of phenolics on performance of the autumnal moth depend on herbivory history of trees. Journal of Chemical Ecology **33**:1160–1176.

Ryals, J. A., U. H. Neuenschwander, M. G. Willits, A. Molina, H. Y. Steiner, and M. D. Hunt. 1996. Systemic acquired resistance. Plant Cell **8**:1809–1819.

Sabelis, M. W., B. P. Afman, and P. J. Slim. 1984. Location of distant spider-mite colonies by *Phytoseiulus persimils*: localization of a kairomone. Pages 431–440 *in* D. A. Griffiths and C. E. Bowman, editors. Acarology VI, vol. 1. Halsted Press, New York.

Sabelis, M. W., and H. E. van de Baan. 1983. Location of distant spider mite colonies by phytoseiid predators: demonstration of specific kairmones emitted by *Tetranychus urticae* and *Panonchus ulmi*. Entomologia Experimentalis et Applicata **33**:303–314.

Sanghera, G. S., S. H. Wani, W. Hussain, and N. B. Singh. 2011. Engineering cold stress tolerance in crop plants. Current Genomics **12**:30–43.

Sapir, Y., A. Shmida, and G. Ne'eman. 2006. Morning floral heat as a reward to the pollinators of the Oncocyclus irises. Oecologia **147**:53–59.

Saracino, A., C. M. D'Alessandro, and M. Borghetti. 2004. Seed colour and post-fire bird predation in a Mediterranean pine forest. Acta Oecologia—International Journal of Ecology **26**:191–196.

Schacter, D. L., D. T. Gilbert, and D. M. Wegner. 2011. Introducing Psychology. Worth, New York.

Schaefer, H. M., D. J. Levey, V. Schaefer, and M. L. Avery. 2006. The role of chromatic and achromatic signals for fruit detection by birds. Behavioral Ecology **17**:784–789.

Schaefer, H. M., and G. D. Ruxton. 2011. Plant-Animal Communication. Oxford University Press, Oxford.

Schaefer, H. M., V. Schaefer, and D. J. Levey. 2004. How plant-animal interactions signal new insights in communication. Trends in Ecology and Evolution **19**:577–584.

Schaefer, H. M., and V. Schmidt. 2004. Detectability and content as opposing signal characteristics in fruits. Proceedings of the Royal Society B **271**:S370–S373.

Schaefer, H. M., V. Schmidt, and H. Winkler. 2003. Testing the defence trade-off hypothesis: how contents of nutrients and secondary compounds affect fruit removal. Oikos **102**:318–328.

Schemske, D. W., J. Agren, and J. Le Corff. 1996. Deceit pollination in the monoecious, neotropical herb *Begonia oaxacana* (Begoniaceae). Pages 292–318 *in* D. G. Lloyd and S. C. H. Barrett, editors. Floral Biology: Studies on Floral Evolution in Animal-Pollinated Plants. Chapman and Hall, New York.

Schenk, H. J. 2006. Root competition: beyond resource depletion. Journal of Ecology **94**:725–739.

Schenk, H. J., S. Espino, C. M. Goedhart, M. Nordenstahl, H. I. M. Cabrera, and C. S. Jones. 2008. Hydraulic integration and shrub growth form linked across continental aridity gradients. Proceedings of the National Academy of Sciences **105**:11248–11253.

Schenk, H. J., and E. W. Seabloom. 2010. Evolutionary ecology of plant signals and toxins: a conceptual framework. Pages 1–19 *in* F. Baluska and V. Ninkovic, editors. Plant Communication from an Ecological Perspective. Springer-Verlag, Berlin.

Schiestl, F. P. 2004. Floral evolution and pollinator mate choice in a sexually deceptive orchid. Journal of Evolutionary Biology **17**:67–75.

Schiestl, F. P. 2005. On the success of a swindle: pollination by deception in orchids. Naturwissenschaften **92**:255–264.

Schittko, U., and I. T. Baldwin. 2003. Constraints to herbivore-induced systemic responses: bidirectional signaling along orthostichies in *Nicotiana attenuata*. Journal of Chemical Ecology **29**:763–770.

Schluter, P. M., and F. P. Schiestl. 2008. Molecular mechanisms of floral mimicry in orchids. Trends in Plant Science **13**:228–235.

Schmidt, V., H. M. Schaefer, and H. Winkler. 2004. Conspicuousness, not colour as foraging cue in plant-animal signalling. Oikos **106**:551–557.

Schmitt, J., J. Niles, and R. D. Wulff. 1992. Norms of reaction of seed traits to maternal environments in *Plantago lanceolata*. American Naturalist **139**:451–466.

Schnee, C., T. G. Kollner, M. Held, T. C. J. Turlings, J. Gershenzon, and J. Degenhardt. 2006. The products of a single maize sesquiterpene synthase form a volatile defense signal that attracts natural enemies of maize herbivores. Proceedings of the National Academy of Sciences **103**:1129–1134.

Schneider, E. L., and J. D. Buchanan. 1980. Morphological studies of the Nymphaeaceae. 11. Floral biology of *Nelumbo pentapetala*. American Journal of Botany **67**:182–193.

Schulze, B., P. Dabrowska, and W. Boland. 2007. Rapid enzymatic isomerization of 12-oxophytodienoic acid in the gut of lepidopteran larvae. ChemBioChem **8**:208–218.

Schupp, E. W., and D. H. Feener. 1991. Phylogeny, life form, and habitat dependance of ant-defended plants in a Panamanian forest. Pages 175–197 *in* D. F. Cutler and C. R. Huxley, editors. Ant-Plant Interactions. Oxford University Press, Oxford.

Schutz, S., B. Weissbecker, U. T. Koch, and H. E. Hummel. 1999. Detection of volatiles released by diseased potato tubers using a biosensor on the basis of intact insect antennae. Biosensors and Bioelectronics **14**:221–228.

Schweiger, P. F., A. D. Robson, and N. J. Barrow. 1995. Root hair length determines beneficial effect of a *Glomus* species on shoot growth of some pasture species. New Phytologist **131**:247–254.

Scott, P. 2008. Physiology and Behaviour of Plants. Wiley, Chichester, UK.

Scott-Phillips, T. G. 2008. Defining biological communication. Journal of Evolutionary Biology **21**:387–395.

Searcy, W. A., and S. Nowicki. 2005. The Evolution of Animal Communication: Reliability and Deception in Signaling Systems. Princeton University Press, Princeton, NJ.

Semchenko, M., K. Zobel, A. Heinemeyer, and M. J. Hutchings. 2008. Foraging for space and avoidance of physical obstructions by plant roots: a comparative study of grasses from contrasting habitats. New Phytologist **179**:1162–1170.

Seymour, R. S. 2004. Dynamics and precision of thermoregulatory responses of eastern skunk cabbage *Symplocarpus foetidus*. Plant, Cell and Environment **27**:1014–1022.

Seymour, R. S., and P. Schultze-Motel. 1996. Thermoregulating lotus flowers. Nature **383**:305.

Seymour, R. S., and P. Schultze-Motel. 1997. Heat-producing flowers. Endeavour **21**:125–129.

Shahbaz, M., M. Ashraf, F. Al-Qurainy, and P. J. C. Harris. 2012. Salt tolerance in selected vegetable crops. Critical Reviews in Plant Sciences **31**:303–320.

Shapiro, A. M. 1981. Egg-mimics of *Streptanthus* (Cruciferae) deter oviposition by *Pieris sisymbrii* (Lepidoptera, Pieridae). Oecologia **48**:142–143.

Shapiro, A. M., and J. E. DeVay. 1987. Hypersensitivity reaction of *Brassica nigra* L. (Cruciferae) kills eggs of Pieris butterflies (Lepidoptera: Pieridae). Oecologia **71**:631–632.

Sharrock, R. A., and T. Clark. 2002. Patterns of expression and normalized levels of the five Arabidopsis phytochromes. Plant Physiology **130**:442–456.

Shelton, A. L. 2005. Within-plant variation in glucosinolate concentrations of *Raphanus sativus* across multiple scales. Journal of Chemical Ecology **31**:1711–1732.

Shemesh, H., A. Arbiv, M. Gersani, O. Ovadia, and A. Novoplansky. 2010. The effects of nutrient dynamics on root patch choice. PLoS One **5**:e10824.

Sherman, W. B., and P. M. Lyrene. 1984. Biannual peaches in the tropics. Fruit Varieties Journal **38**:37–39.

Shiojiri, K., K. Kishimoto, R. Ozawa, S. Kugimiya, S. Urashimo, G. Arimura, J. Horiuchi, T. Nishioka, K. Matsui, and J. Takabayashi. 2006. Changing green leaf volatile biosynthesis in plants: an approach for improving plant resistance against both herbivores and pathogens. Proceedings of the National Academy of Sciences **103**:16672–16676.

Shirley, B. W. 1996. Flavonoid biosynthesis: "new" functions for an "old" pathway. Trends in Plant Science **1**:377–382.

Shulaev, V., P. Silverman, and I. Raskin. 1997. Airborne signalling by methly salicylate in plant pathogen resistance. Nature **385**:718–721.

Sih, A., and A. M. Bell. 2008. Insights for behavioral ecology from behavioral syndromes. Advances in the Study of Behavior **38**:227–281.

Silvertown, J. 1998. Plant phenotypic plasticity and non-cognitive behaviour. Trends in Ecology and Evolution **13**:255–256.

Silvertown, J., and D. M. Gordon. 1989. A framework for plant behavior. Annual Review of Ecology and Systematics **20**:349–366.

Simard, S. W., K. J. Beiler, M. A. Bingham, J. R. Deslippec, L. J. Philip, and F. P. Testee. 2012. Mycorrhizal networks: mechanisms, ecology and modelling. Fungal Biology Reviews **26**:39–60.

Simons, P. 1992. The Action Plant: Movement and Nervous Behaviour in Plants. Blackwell, Oxford.

Skogsmyr, I., and A. Lankinen. 2002. Sexual selection: an evolutionary force in plants. Biological Reviews **77**:537–562.

Slansky, F. 1978. Utilization of energy and nitrogen by larvae of the imported cabbageworm, *Pieris rapae*, as affected by parasitism by *Apanteles glomeratus*. Environmental Entomology **7**:178–185.

Slaughter, A., X. Daniel, V. Flors, E. Luna, B. Hohn, and B. Mauch-Mani. 2012. Descendents of primed Arabidopsis plants exhibit resistance to biotic stress. Plant Physiology **158**:835–843.

Smith, A. M., and M. Stitt. 2007. Coordination of carbon supply and plant growth. Plant, Cell and Environment **30**:1126–1149.

Smith, F. A., E. J. Grace, and S. E. Smith. 2009. More than a carbon economy: nutrient trade and ecological sustainability in facultative arbuscular mycorrhizal symbioses. New Phytologist **182**:347–358.

Smith, H. 1982. Light quality, photoperception, and plant strategy. Annual Review of Plant Physiology and Plant Molecular Biology **33**:481–518.

Smith, H. 1992. The ecological functions of the phytochrome family: clues to a transgenic program of crop improvement. Photochemistry and Photobiology **56**:815–822.

Smith, H. 2000. Phytochromes and light signal perception by plants—an emerging synthesis. Nature **407**:585–591.

Smith, L. L., J. Lanza, and G. C. Smith. 1990. Amino acid concentrations in extrafloral nectar of *Impatiens sultani* increase after simulated herbivory. Ecology **71**:107–115.

Smith, S. E., and D. J. Read. 2008. Mycorrhizal Symbiosis, 3rd ed. Academic Press, New York.

Smith, W. J. 1977. The Behavior of Communication. Harvard University Press, Cambridge, MA.

Smithson, A. 2002. The consequences of rewardlessness in orchids: reward-supplementation experiments with *Anacamptis morio* (Orchidaceae). American Journal of Botany **89**:1579–1587.

Smithson, A., and L. D. B. Gigord. 2003. The evolution of empty flowers revisited. American Naturalist **161**:537–552.

Snow, D. W. 1971. Evolutionary aspects of fruit-eating by birds. Ibis **113**:194–202.

Soltau, U., S. Doetterl, and S. Liede-Schumann. 2009. Leaf variegation in *Caladium steudneriifolium* (Araceae): a case of mimicry? Evolutionary Ecology **23**:503–512.

Somanathan, H., R. M. Borges, E. J. Warrant, and A. Kelber. 2008. Visual ecology of Indian carpenter bees I: light intensities and flight activity. Journal of Comparative Physiology A **194**:97–107.

Song, Y. Y., R. S. Zeng, J. F. Xu, J. Li, X. Shen, and W. G. Yihdego. 2010. Interplant communication of tomato plants through underground common mycorrhizal networks. PLoS One **5**:e13324.

Sorensen, A. E. 1986. Seed dispersal by adhesion. Annual Review of Ecology and Systematics **17**:443–463.

Spaethe, J., J. Tautz, and L. Chittka. 2001. Visual constraints in foraging bumblebees: flower size and color affect search time and flight behavior. Proceedings of the National Academy of Sciences **98**:3898–3903.

Speirs, J., E. Lee, K. Holt, K. Yong-Duk, N. S. Scott, B. Loveys, and W. Schuch. 1998. Genetic manipulation of alcohol dehydrogenase levels in ripening tomato fruit affects the balance of some flavor aldehydes and alcohols. Plant Physiology **117**:1047–1058.

Spoel, S. H., A. Koornneef, S. M. C. Claessens, J. P. Korzelius, J. A. Van Pelt, M. J. Mueller, A. J. Buchala, J.-P. Metraux, R. Brown, K. Kazan, L. C. Van Loon, X. Dong, and C. M. J. Pieterse. 2003. NPR1 modulates cross-talk between salicylate- and jasmonate-dependent pathways through a novel function in the cytosol. Plant Cell **15**:760–770.

Sprengel, C. K. 1793. Das entdeckte Geheimniss der Natur im Bau und in der Befruchtung der Blumen. Vieweg, Berlin.

Sprengel, C. K. 1996. Discovery of the secret of nature in the structure and fertilization of flowers. Pages 3–43 *in* D. G. Lloyd and S. C. H. Barrett, editors. Floral Biology: Studies on Floral Evolution in Animal-Pollinated Plants. Chapman and Hall, New York.

Stanton, M. L., T. M. Palmer, T. P. Young, A. Evans, and M. L. Turner. 1999. Sterilization and canopy modification of a swollen thorn acacia tree by a plant-ant. Nature **401**:576–581.

Stephenson, A. G. 1982. The role of extrafloral nectaries of *Catalpa speciosa* in limiting herbivory and increasing fruit production. Ecology **63**:663–669.

Stephenson, A. G., S. E. Travers, J. I. Mena-Ali, and J. A. Winsor. 2003. Pollen performance before and during the autotrophic-heterotrophic transition of pollen tube growth. Philosophical Transactions of the Royal Society of London B **358**:1009–1017.

Sticher, L., B. Mauch-Mani, and J.-P. Metraux. 1997. Systemic acquired resistance. Annual Review of Phytopathology **35**:235–270.

Stork, W., C. Diezel, R. Halitschke, I. Galis, and I. T. Baldwin. 2009. An ecological analysis of the herbivory-elicited JA burst and its metabolism: plant memory processes and predictions of the moving target model. PLoS One **4**:e4697.

Stout, M. J., J. S. Thaler, and B. P. H. J. Thomma. 2006. Plant-mediated interactions between pathogenic microorganisms and herbivorous arthropods. Annual Review of Entomology **51**:663–689.

Stout, M. J., J. Workman, and S. S. Duffey. 1996. Identity, spatial distribution, and variability of induced chemical responses in tomato plants. Entomologia Experimentalis et Applicata **79**:255–271.

Strauss, S. D. 2002. The Big Idea: How Business Innovators Get Great Ideas to Market. Dearborn Trade Publishing, Chicago.

Strauss, S. Y., D. H. Siemens, M. B. Decher, and T. Mitchell-Olds. 1999. Ecological costs of plant resistance to herbivores in the currency of pollination. Evolution **53**:1105–1113.

Strauss, S. Y., and J. B. Whittall. 2006. Non-pollinator agents of selection on floral traits. Pages 120–138 *in* L. D. Harder and S. C. H. Barrett, editors. Ecology and Evolution of Flowers. Oxford University Press, Oxford.

Stuefer, J. F., H. de Kroon, and H. J. During. 1996. Exploitation of environmental heterogeneity by spatial division of labour in a clonal plant. Functional Ecology **10**:328–334.

Taiz, L., and E. Zeiger. 2002. Plant Physiology, 3rd ed. Sinauer, Sunderland, MA.

Taiz, L., and E. Zeiger. 2010. Plant Physiology, 5th ed. Sinauer, Sunderland, MA.

Takabayashi, J., M. W. Sabelis, A. Janssen, K. Shiojiri, and M. van Wijk. 2006. Can plants betray the presence of multiple herbivore species to predators and parasitoids? The role of learning in phytochemical information networks. Ecological Research **21**:3–8.

Takabayashi, J., S. Takahashi, M. Dicke, and M. A. Posthumus. 1995. Developmental stage of herbivore *Pseudaletia separata* affects herbivore-induced synomone by corn plants. Journal of Chemical Ecology **21**:273–287.

Takayama, S., and A. Isogai. 2005. Self-incompatibiity in plants. Annual Review of Plant Biology **56**:467–489.

Takeuchi, H., and T. Higashiyama. 2011. Attraction of tip-growing pollen tubes by the female gametophyte. Current Opinion in Plant Biology **14**:614–621.

Tamogami, S., R. Ralkwal, and G. K. Agrawal. 2008. Interplant communication: airborne methyl jasmonate is essentially converted into JA and JA-Ile activating jasmonate signaling pathway and VOCs emission. Biochemical and Biophysical Research Communications **376**:723–727.

Tang, H. Q., J. Hu, L. Yang, and R. X. Tan. 2000. Terpenoids and flavonoids from *Artemisia* species. Planta Medica **66**:391–393.

Tang, W. H., D. Kelley, I. Ezcurra, R. Cotter, and S. McCormick. 2004. LeSTIG1, an extracellular binding partner for the pollen receptor kinases LePRK1 and LePRK2, promotes pollen tube growth in vitro. Plant Journal **39**:343–353.

Taylor, P. E., G. Card, J. House, M. H. Dickinson, and R. C. Flagan. 2006. High-speed pollen release in the white mulberry tree, *Morus alba* L. Sexual plant reproduction **19**:19–24.

Telewski, F. W. 2006. A unified hypothesis of mechanoperception in plants. American Journal of Botany **93**:1466–1476.

Terborgh, J. 2012. Enemies maintain hyperdiverse tropical forests. American Naturalist **179**:303–314.

ter Horst, C. and J. A. Lau. 2012. Direct and indirect transgenerational effects alter plant-herbivore interactions. Evolutionary Ecology **26**:1469–1480.

Terry, I., C. J. Moore, G. H. Walter, P. I. Forster, R. B. Roemer, J. D. Donaldson, and P. J. Machin. 2004. Association of cone thermogenesis and volatiles with pollinator specificity in Macrozamia cycads. Plant Systematics and Evolution **243**:223–247.

Terry, I., G. H. Walter, C. Moore, R. Roemer, and C. Hull. 2007. Odor-mediated push-pull pollination in cycads. Science **318**:70.

Tewksbury, J. J., D. J. Levey, M. Huizinga, D. C. Haak, and A. Traveset. 2008. Costs and benefits of capsaicin-mediated oontrol of gut retention in dispersers of wild chilies. Ecology **89**:107–117.

Tewksbury, J. J., and G. P. Nabhan. 2001. Seed dispersal: directed deterrence by capsaicin in chilies. Nature **412**:403–404.

Tewksbury, J. J., K. M. Reagan, N. J. Machnicki, T. A. Carlo, D. C. Haak, A. L. C. Penaloza, and D. J. Levey. 2008. Evolutionary ecology of pungency in wild chilies. Proceedings of the National Academy of Sciences **105**:11808–11811.

Thakar, J. D., V. Kunte, A. K. Chauhan, A. V. Watve, and M. G. Watve. 2003. Nectarless flowers: ecological correlates and evolutionary stability. Oecologia **136**:565–570.

Thaler, J. S. 1999a. Jasmonate-inducible plant defences cause increased parasitism of herbivores. Nature **399**:686–688.

Thaler, J. S. 1999b. Induced resistance in agricultural crops: effects of jasmonic acid on herbivory and yield in tomato plants. Environmental Entomology **28**:30–37.

Thaler, J. S., M. A. Farag, P. W. Pare, and M. Dicke. 2002. Jasmonate-deficient plants have reduced direct and indirect defences against herbivores. Ecology Letters **5**:764–774.

Thaler, J. S., P. T. Humphrey, and N. K. Whiteman. 2012. Evolution of jasmonate and salicylate signal crosstalk. Trends in Plant Science **17**:260–270.

Thaler, J. S., M. J. Stout, R. Karban, and S. S. Duffey. 1996. Exogenous jasmonates simulate insect wounding in tomato plants (*Lycopersicon esculentum*) in the laboratory and field. Journal of Chemical Ecology **22**:1767–1781.

Theimer, T. C., and G. C. Bateman. 1992. Patterns of prickly-pear herbivory by collared peccaries. Journal of Wildlife Management **56**:234–240.

Thines, B., L. Katsir, M. Melotto, Y. Niu, A. Mandaokar, G. Liu, K. Nomura, S. Y. He, G. A. Howe, and J. Browse. 2007. JAZ repressor proteins are targets of the SCFCO11 complex during jasmonate signalling. Nature **448**:661–665.

Thomashow, M. F. 1999. Plant cold acclimation: freezing tolerance genes and regulatory mechanisms. Annual Review of Plant Physiology and Plant Molecular Biology **50**:571–599.

Thomson, J. D., B. J. Andrews, and R. C. Plowright. 1982. The effect of a foreign pollen on ovule development in *Diervilla lonicera* (Caprifoliaceae). New Phytologist **90**:777–783.

Thorpe, A. S., G. C. Thelan, A. Diaconu, and R. M. Callaway. 2009. Root exudate is alle-
lopathic in invaded community but not in native community: field evidence for the
novel weapons hypothesis. Journal of Ecology **97**:641–645.

Tinbergen, N. 1955. The Study of Instinct. Clarendon, Oxford.

Tompkins, P., and C. Bird. 1973. The Secret Lives of Plants. Harper and Rowe, New York.

Ton, J., M. D'Alessandro, V. Jourdie, G. Jakab, D. Karlen, M. Held, B. Mauch-Mani, and
T. C. J. Turlings. 2007. Priming by airborne signals boosts direct and indirect resistance in maize. Plant Journal **49**:16–26.

Torres, M. A. 2010. ROS in biotic interactions. Physiologia Plantarum **138**:414–429.

Trewavas, A. 2009. What is plant behaviour? Plant, Cell and Environment **32**:606–616.

Tripp, E. A., and P. S. Janos. 2008. Is floral specialization an evolutionary dead-end? Pollination system transitions in Ruellia (Acanthaceae). Evolution **62**:1712–1736.

Truitt, C. L., H. X. Wei, and P. W. Pare. 2004. A plasma membrane protein from *Zea mays*
binds with the herbivore elicitor volicitin. Plant Cell **16**:523–532.

Tscharntke, T., S. Thiessen, R. Dolch, and W. Boland. 2001. Herbivory, induced resistance, and interplant signal transfer in *Alnus glutinosa*. Biochemical Systematics
and Ecology **29**:1025–1047.

Tufto, J. 2000. The evolution of plasticity and nonplastic spatial and temporal adapa-
tions in the presence of imperfect environmental cues. American Naturalist
156:121–130.

Tuomi, J., P. Niemela, M. Rousi, S. Siren, and T. Vuorisalo. 1988. Induced accumulation
of foliage phenols in mountain birch: branch response to defoliation. American
Naturalist **132**:602–608.

Turkington, R., R. S. Hamilton, and C. Gliddon. 1991. Within-population variation in
localized and integrated responses of *Trifolium repens* to biotically patchy environments. Oecologia **86**:183–192.

Turlings, T. C. J., I. Hiltpold, and S. Rasmann. 2012. The importance of root-produced
volatiles as foraging cues for entomopathogenic nematodes. Plant and Soil
358:47–56.

Turlings, T. C. J., U. B. Lengwiler, M. L. Bernasconi, and D. Wechsler. 1998. Timing of
induced volatile emission in maize seedlings. Planta **207**:146–152.

Turlings, T. C. J., and J. H. Tumlinson. 1992. Systemic release of chemical signals by
herbivore-injured corn. Proceedings of the National Academy of Sciences **89**:8399–
8402.

Turlings, T. C. J., J. H. Tumlinson, F. J. Eller, and W. J. Lewis. 1991. Larval-damaged
plants—sources of volatile synomones that guide the parasitoid *Cotesia margini-
ventris* to the microhabitat of its hosts. Entomologia Experimentalis et Applicata
58:75–82.

Turlings, T. C. J., J. H. Tumlinson, and W. J. Lewis. 1990. Exploitation of herbivore-
induced plant odors by host-seeking parasitic wasps. Science **250**:1251–1253.

Turlings, T. C. J., and F. Wackers. 2004. Recruitment of predators and parasitoids by
herbivore-injured plants. Pages 21–75 in R. T. Carde and J. G. Millar, editors. Advances in Insect Chemical Ecology. Cambridge University Press, Cambridge.

Turlings, T. C. J., F. Wackers, L. E. M. Vet, W. J. Lewis, and J. H. Tumlinson. 1993. Learning of host-finding cues by hymenopterous parasitoids. Pages 51–78 in D. R. Papaj
and A. C. Lewis, editors. Insect Learning: Ecology and Evolutionary Perspectives.
Chapman and Hall, New York.

Underwood, N. 1998. The timing of induced resistance and induced susceptibility in the
soybean–Mexican bean beetle system. Oecologia **114**:376–381.

Valido, A., H. M. Schaefer, and P. Jordano. 2011. Colour, design and reward: phenotypic integration of fleshy fruit displays. Journal of Evolutionary Biology **24**:751–760.

Vallad, G. E., and R. M. Goodman. 2004. Systemic acquired resistance and induced systemic resistance in conventional agriculture. Crop Science **44**:1920–1934.

Vancanneyt, G., C. Sanz, T. Farmaki, M. Paneque, F. Ortego, P. Castanera, and J. J. Sanchez-Serrano. 2001. Hydroperoxide lyase depletion in transgenic potato plants leads to an increase in aphid performance. Proceedings of the National Academy of Sciences **98**:8139–8144.

Vandenbussche, F., R. Pierik, F. F. Millenaar, L. A. C. J. Voesenek, and V. D. Straeten. 2005. Reaching out of the shade. Current Opinion in Plant Biology **8**:462–468.

Van der Ent, S., M. Van Hulten, M. J. Pozo, T. Czechowski, M. K. Udvardi, C. M. J. Pieterse, and J. Ton. 2009a. Priming of plant innate immunity by rhizobacteria and beta-aminobutyric acid: differences and similarities in regulation. New Phytologist **183**:419–431.

Van der Ent, S., S. C. M. Van Wees, and C. M. J. Pieterse. 2009b. Jasmonate signaling in plant interactions with resistance-inducing beneficial microbes. Phytochemistry **70**:1581–1588.

van der Heijden, M. G. A., J. N. Klironomos, M. Ursic, P. Moutoglis, R. Streitwolf-Engel, T. Boller, A. Wiemken, and I. R. Sanders. 1998. Mycorrhizal fungal diversity determines plant biodiversity, ecosystem variability and productivity. Nature **396**:69–72.

Vander Wall, S. B. 2001. The evolutionary ecology of nut dispersal. Botanical Review **67**:74–117.

van Doorn, W. G. 1997. Effects of pollination on floral attraction and longevity. Journal of Experimental Botany **48**:1615–1622.

van Hulten, M., M. Pelser, L. C. Van Loon, C. M. J. Pieterse, and J. Ton. 2006. Costs and benefits of priming for defense in Arabidopsis. Proceedings of the National Academy of Sciences **103**:5602–5607.

van Loon, L. C., P. A. H. M. Bakker, and C. M. J. Pieterse. 1998. Systemic resistance induced by rhizosphere bacteria. Annual Review of Phytopathology **36**:453–483.

van Loon, L. C., M. Rep, and C. M. J. Pieterse. 2006a. Significance of inducible defense-related proteins in infected plants. Annual Review of Phytopathology **44**:135–162.

van Loon, L. C., B. P. J. Geraats, and H. J. M. Linthorst. 2006b. Ethylene as a modulator of disease resistance in plants. Trends in Plant Science **11**:184–191.

Vannette, R. L., M.-P. L. Gauthier, and T. Fukami. 2013. Nectar bacteria, but not yeast, weaken a plant-pollinator mutualism. Proceedings of the Royal Society B **280**:20122601.

Vannini, C., M. Campa, M. Iriti, A. Genda, F. Faoro, S. Carravieri, G. L. Rotino, M. Rossoni, A. Spinardi, and M. Bracale. 2007. Evaluation of transgenic tomato plants ectopically expressing the rice Osmyb4 gene. Plant Science **173**:231–239.

Van Oosten, V. R., N. Bodenhausen, P. Reymond, J. A. Van Pelt, L. C. Van Loon, M. Dicke, and C. M. J. Pieterse. 2008. Differential effectiveness of microbially induced resistance against herbivorous insects in Arabidopsis. Molecular Plant-Microbe Interactions **21**:919–930.

Van Tienderen, P. H. 1991. Evolution of generalists and specialists in spatially heterogeneous environments. Evolution **45**:1317–1331.

Van Tol, R. W. H. M., A. T. C. van der Sommen, M. I. C. Boff, J. van Bezooijen, M. W. Sabelis, and P. H. Smits. 2001. Plants protect their roots by alerting the enemies of grubs. Ecology Letters **4**:292–294.

van Vuuren, D. P., O. E. Sala, and H. M. Pereira. 2006. The future of vascular plant diversity under four global scenarios. Ecology and Society **11**:e25.

Varassin, I. G., J. R. Trigo, and M. Sazima. 2001. The role of nectar production, flower pigments and odour in the pollination of four species of *Passiflora* (Passifloraceae) in south-eastern Brazil. Botanical Journal of the Linnean Society **136**:139–152.

Vazquez-Yanes, C., and H. Smith. 1982. Phytochrome control of seed-germination in the tropical rain-forest pioneer trees *Cecropia obtusifolia* and *Piper auritum* and its ecological significance. New Phytologist **92**:477–485.

Verhoeven, K. J. F., and T. P. van Gurp. 2012. Transgenerational effects of stress exposure on offspring phenotypes in apomictic dandelions. PLoS One **7**:e38605.

Vet, L. E. M., and M. Dicke. 1992. Ecology of infochemical use by natural enemies in a tritrophic context. Annual Review of Entomology **37**:141–172.

Via, S. 1987. Genetic constraints on the evolution of phenotypic plasticity. Pages 47–69 *in* V. Loeschcke, editor. Genetic Constraints on Adaptive Evolution. Springer-Verlag, Berlin.

Vickers, C. E., J. Gershenzon, M. T. Lerdau, and F. Lotero. 2009. A unified mechanism of action for volatile isoprenoids in plant abiotic stress. Nature Chemical Biology **5**:283–291.

Visser, J. H. 1986. Host odor perception in phytophagous insects. Annual Review of Entomology **31**:121–144.

Viswanathan, D. V., and J. S. Thaler. 2004. Plant vascular architecture and within-plant spatial patterns in resource quality following herbivory. Journal of Chemical Ecology **30**:531–543.

Viswanathan, D. V., O. A. Lifchits, and J. S. Thaler. 2007. Consequences of sequential attack for resistance to herbivores when plants have specific induced responses. Oikos **116**:1389-1399.

Vivas, M., J. A. Martin, L. Gil, and A. Solla. 2012. Evaluating methyl jasmonate for induction of resistance to *Fusarium oxysporum*, *F. circinatum* and *Ophiostoma novo-ulmi*. Forest Systems **21**:289-299.

Vlot, A. C., P.-P. Liu, R. K. Cameron, S.-W. Park, Y. Yang, D. Kumar, F. Zhou, T. Padukkavidana, C. Gustafsson, E. Pichersky, and D. F. Klessig. 2008. Identification of likely orthologs of tobacco salicylic acid–binding protein 2 and their role in systemic acquired resistance in *Arabidopsis thaliana*. Plant Journal **56**:445–456.

Vogt-Eisele, A. K., K. Keber, M. A. Sherkhili, G. Vielhaber, J. Panten, G. Gisselmann, and H. Hatt. 2007. Monoterpenoid agonists of TRPV3. British Journal of Pharmacology **151**:530–540.

Volkov, A. G., T. Adesina, V. S. Markin, and E. Jovanov. 2008. Kinetics and mechanism of *Dionaea muscipula* trap closing. Plant Physiology **146**:696-702.

Volkov, A. G., J. C. Foster, T. A. Ashby, R. K. Walker, J. A. Johnson, and V. S. Markin. 2010. *Mimosa pudica*: electrical and mechanical stimulation of plant movements. Plant, Cell and Environment **33**:163–173.

von Dahl, C. C., and I. T. Baldwin. 2007. Deciphering the role of ethylene in plant-herbivore interactions. Journal of Plant Growth Regulation **26**:201–209.

von Frisch, K. 1919. Über den Geruchssinn der Bienen und seine blütenbiologische Bedeutung. Zoologische Jahrbücher (Physiologie) **37**:1–238.

von Merey, G., N. Veyrat, G. Mahuku, R. L. Valdez, T. C. J. Turlings, and M. D'Alessandro. 2011. Dispensing synthetic green leaf volatiles in maize fields increases the release of sesquiterpenes by the plants, but has little effect on the attraction of pest and beneficial insects. Phytochemistry **72**:1838-1847.

von Rad, U., M. J. Mueller, and J. Durner. 2005. Evaluation of natural and synthetic stimulants of plant immunity by microarray technology. New Phytologist **165**:191–202.

Vuorinen, T., A.-M. Nerg, M. A. Ibrahim, G. V. P. Reddy, and J. K. Holopainen. 2004. Emission of *Plutella xylostella*-induced compounds from cabbages grown at elevated CO_2 and orientation behavior of the natural enemies. Plant Physiology **135**:1984–1992.

Wagner, A. M., S. Krab, M. J. Wagner, and A. L. Moore. 2008. Regulation of thermogenesis in flowering Araceae: the role of the alternative oxidase. Biochimica et Biophysica Acta **1777**:993–1000.

Wahaj, S. A., D. J. Levey, A. K. Sanders, and M. L. Cipollini. 1998. Control of gut retention time by secondary metabolites in ripe Solanum fruits. Ecology **79**:2309–2319.

Wainwright, M. 1989. Moulds in ancient and more recent medicine. Mycologist **3**:21–23.

Waisel, Y., N. Liphschitz, and Z. Kuller. 1972. Patterns of water movement in trees and shrubs. Ecology **53**:520–523.

Waite, S. 1994. Field evidence of plastic growth responses to habitat heterogeneity in the clonal herb *Ranunculus repens*. Ecological Research **9**:311–316.

Walter, D. E. 1996. Living on leaves: mites, tomenta, and leaf domatia. Annual Review of Entomology **41**:101–114.

Wang, W., J. J. Esch, S. H. Shiu, H. Agula, B. M. Binder, C. Chang, S. E. Patterson, and A. B. Bleecker. 2006. Identification of important regions for ethylene binding and signaling in the transmembrane domain of the ETR1 ethylene receptor of Arabidopsis. Plant Cell **18**:3429–3442.

Wargent, J. J., D. A. Pickup, N. D. Paul, and M. R. Roberts. 2013. Reduction of photosynthetic sensitivity in response to abiotic stress in tomato is mediated by a new generation plant activator. BMC Plant Biology **13**:108.

Waser, N. M. 1986. Flower constancy—definition, cause, and measurement. American Naturalist **127**:593–603.

Waser, N. M. 2001. Pollinator behavior and plant speciation: looking beyond the "ethological isolation" paradigm. Pages 318–335 *in* L. Chittka and J. D. Thomson, editors. Cognitive Ecology of Pollination: Animal Behavior and Floral Evolution. Cambridge University Press, Cambridge.

Waser, N. M., L. Chittka, M. V. Price, N. M. Williams, and J. Ollerton. 1996. Generalization in pollination systems, and why it matters. Ecology **77**:1043–1060.

Waser, N. M., and M. V. Price. 1981. Pollinator choice and stabilizing selection for flower color in *Delphinium nelsonii*. Evolution **35**:376–390.

Watson, M. A., and B. B. Casper. 1984. Morphogenetic constraints on patterns of carbon distribution in plants. Annual Review of Ecology and Systematics **15**:233–258.

Weaver, J. E. 1926. Root Development in Field Crops. McGraw Hill, New York.

Webster, B., T. Bruce, J. Pickett, and J. Hardie. 2010. Volatiles functioning as host cues in a blend become nonhost cues when presented alone to the black bean aphid. Animal Behaviour **79**:451–457.

Weinberger, P., and U. Graefe. 1973. The effect of variable-frequency sounds on plant growth. Canadian Journal of Botany **51**:1851–1856.

Weinig, C., and L. F. Delph. 2001. Phenotypic plasticity early in life constrains developmental responses later. Evolution **55**:930-936.

Weinig, C., J. Johnston, Z. M. German, and L. M. Demink. 2006. Local and global costs of adaptive plasticity to density in *Arabidopsis thaliana*. American Naturalist **167**:826–836.

Weir, T. L., H. P. Bais, V. J. Stull, R. M. Callaway, G. C. Thelan, W. M. Ridenour, S. Bhamidi, F. R. Stermitz, and J. M. Vivianco. 2006. Oxalate contributes to the resistance of *Gaillardia grandiflora* and *Lupinus sericeas* to a phytotoxin produced by *Centaurea maculosa*. Planta **223**:785–795.

Weiss, M. R. 1995. Floral color change: a widespread functional convergence. American Journal of Botany **82**:167–185.

Wenny, D. G., and D. J. Levey. 1998. Directed seed dispersal by bellbirds in a tropical cloud forest. Proceedings of the National Academy of Sciences **95**:6204–6207.

Went, F. W. 1926. On growth-accelerating substances in the coleoptile of *Avena sativa*. Proceedings of the Koninklijke Nederlandse Akademie van Wetenschappen **30**:10–19.

Wheeler, M. J., S. Vatovec, and V. E. Franklin-Tong. 2010. The pollen S-determinant in Papaver: comparisons with known plant receptors and protein ligand partners. Journal of Experimental Botany **61**:2015–2025.

White, J. 1979. The plant as a metapopulation. Annual Review of Ecology and Systematics **10**:109–145.

White, J. 1984. Plant metamerism. Pages 15–47 *in* R. Dirzo and J. Sarukhan, editors. Perspectives in Plant Population Ecology. Sinauer, Sunderland, MA.

Whitham, T. G. 1983. Host manipulation of parasites: within-plant variation as a defense against rapidly evolving pests. Pages 15–41 *in* R. F. Denno and M. S. McClure, editors. Variable Plants and Herbivores in Natural and Managed Systems. Academic Press, New York.

Wickler, W. 1968. Mimicry in Plants and Animals (translated by R. D. Martin). McGraw-Hill, New York.

Wiens, D. 1978. Mimicry in plants. Evolutionary Biology **11**:365–403.

Wildon, D. C., J. F. Thain, P. E. H. Minchin, I. R. Gubb, A. J. Reilly, Y. D. Skipper, H. M. Doherty, P. J. O'Donnell, and D. J. Bowles. 1992. Electrical signaling and systemic proteinase-inhibitor induction in the wounded plant. Nature **360**:62–65.

Wiley, R. H. 1983. The evolution of communication: information and manipulation. Pages 156–189 *in* T. R. Halliday and P. J. B. Slater, editors. Communication. Freeman, New York.

Williams, G. C. 1966. Adaptation and Natural Selection. Princeton University Press, Princeton, N.J.

Williams, K. S., and J. H. Myers. 1984. Previous herbivore attack of red alder may improve food quality for fall webworm larvae. Oecologia **63**:166–170.

Williams, N. M. 2003. Use of novel pollen species by specialist and generalist solitary bees (Hymenoptera : Megachilidae). Oecologia **134**:228–237.

Williamson, C. E. 1950. Ethylene, a metabolic product of diseased or injured plants. Phytopathology **40**:205–208.

Willmer, P. 2011. Pollination and Floral Ecology. Princeton University Press, Princeton, NJ.

Willmer, P. G., C. V. Nuttman, N. E. Raine, G. N. Stone, J. G. Pattrick, K. Henson, P. Stillman, L. McIlroy, S. G. Potts, and J. T. Knudsen. 2009. Floral volatiles controlling ant behaviour. Functional Ecology **23**:888–900.

Willson, M. F. 1991. Foliar mites in the eastern deciduous forest. American Midland Naturalist **126**:111–117.

Willson, M. F., D. A. Graff, and C. J. Whelan. 1990. Color preferences of frugivorous birds in relation to the colors of fleshy fruits. Condor **92**:545–555.

Willson, M. F., and C. J. Whelan. 1990. The evolution of fruit color in fleshy-fruited plants. American Naturalist **136**:790–809.

Wilson, E. O. 1975. Sociobiology. Belknap Press, Cambridge, MA.

Wilson, P., M. C. Castellanos, J. N. Hogue, J. N. Thompson, and W. S. Armbruster. 2004. A multivariate search for pollination syndromes among Penstamons. Oikos **104**:345–361.

Wisniewska, J., J. Xu, D. Seifertova, P. B. Brewer, K. Ruzicka, I. Blilou, D. Rouquie, E. Benkova, B. Scheres, and J. Friml. 2006. Polar PIN localization directs auxin flow in plants. Science **312**:883.

Wolf, M., G. S. van Doorn, O. Leimar, and F. J. Weissing. 2007. Life-history trade-offs favour the evolution of animal personalities. Nature **447**:581–584.

Wolff, T. 1950. Pollination and fertilization of the fly Ophrys, *Ophrys insectifera* L. in Allindelille Fredskov, Denmark. Oikos **2**:20–59.

Wolken, J. J. 1995. Light Detectors, Photoreceptors, and Imaging Systems in Nature. Oxford University Press, Oxford.

Worrall, D., G. H. Holroyd, J. P. Moore, M. Glowacz, P. Croft, J. E. Taylor, N. D. Paul, and M. R. Roberts. 2012. Treating seeds with activators of plant defence generates long-lasting priming of resistance to pests and pathogens. New Phytologist **193**:770–778.

Wright, G. A., A. Lutmerding, N. Dudareva, and B. H. Smith. 2005. Intensity and the ratios of compounds in the scent of snapdragon flowers affect scent discrimination by honeybees (*Apis mellifera*). Journal of Comparative Physiology A **191**:105–114.

Wright, G. A., and F. P. Schiestl. 2009. The evolution of floral scent: the influence of olfactory learning by insect pollinators on the honest signalling of floral rewards. Functional Ecology **23**:841–851.

Wu, J., and I. T. Baldwin. 2009. Herbivory-induced signalling in plants: perception and action. Plant, Cell and Environment **32**:1161–1174.

Xu, H. X., N. T. Blair, and D. E. Clapham. 2005. Camphor activates and strongly desensitizes the transient receptor potential vanilloid subtype 1 channel in a vanilloid-independent mechanism. Journal of Neuroscience **25**:8924–8937.

Yan, Y., S. Stolz, A. Chetelat, P. Reymond, M. Pagni, L. Dubugnon, and E. E. Farmer. 2007. A downstream mediator in the growth repression limb of the jasmonate pathway. Plant Cell **19**:2470–2483.

Yarmolinsky, D. A., C. S. Zuker, and N. J. P. Ryba. 2009. Common sense about taste: from mammals to insects. Cell **139**:234–244.

Yasuda, M., A. Ishikawa, Y. Jikumaru, M. Seki, T. Umezawa, T. Asami, A. Maruyama-Nakshita, T. Kudo, K. Shinozaki, S. Yoshida, and H. Nakashita. 2008. Antagonistic interaction between systemic acquired resistance and the abscisic acid–mediated abiotic stress response in Arabidopsis. Plant Cell **20**:1678–1692.

Yi, H.-S., M. Heil, R. M. Adame-Alvarez, D. Ballhorn, and C.-M. Ryu. 2009. Airborne induction and priming of plant defenses against a bacterial pathogen. Plant Physiology **151**:2152-2161.

Yokawa, K., T. Kagenishi, and F. Baluska. 2013. Root photomorphogenesis in laboratory-maintained Arabidopsis seedlings. Trends in Plant Science **18**:117–119.

Young, T. P., and S. P. Hubbell. 1991. Crown symmetry, treefalls, and repeat disturbance of broad-leaved forest gaps. Ecology **72**:1464–1471.

Young, T. P., and B. D. Okello. 1998. Relaxation of an induced defense after exclusion of herbivores: spines on *Acacia drepanolobium*. Oecologia **115**:508–513.

Young, T. P., M. L. Stanton, and C. E. Christian. 2003. Effects of natural and simulated herbivory on spine lengths of *Acacia drepanolobium* in Kenya. Oikos **101**:171–179.

Zahavi, A. 1975. Mate selection: a selection for a handicap. Journal of Theoretical Biology **53**:205–214.

Zahavi, A., and A. Zahavi. 1997. The Handicap Principle: A Missing Piece of Darwin's Puzzle. Oxford University Press, New York.

Zamioudis, C., and C. M. J. Pieterse. 2012. Modulation of host immunity by beneficial microbes. Molecular Plant-Microbe Interactions **25**:139–150.

Zanne, A. E., K. Sweeney, M. Sharma, and C. M. Orians. 2006. Patterns and consequences of differential vascular sectoriality in 18 temperate tree and shrub species. Functional Ecology **20**:200–206.

Zebelo, S. A., and M. Maffei. 2012. Signal transduction in plant-insect interactions: from membrane potential variations to metabolomics. Pages 143-172 *in* A. G. Volkov, editor. Plant Electrophysiology. Springer-Verlag, Berlin.

Zehnder, G. W., J. W. Kloepper, C. G. Yao, and G. Wei. 1997. Induction of systemic resistance in cucumber against cucumber beetles (Coleoptera: Chrysomelidae) by plant growth–promoting rhizobacteria. Journal of Economic Entomology **90**:391–396.

Zehnder, G. W., J. F. Murphy, E. J. Sikora, and J. W. Kloepper. 2001. Application of rhizobacteria for induced resistance. European Journal of Plant Pathology **107**:39050.

Zhang, H. M., A. Jennings, P. W. Barlow, and B. G. Forde. 1999. Dual pathways for regulation of root branching by nitrate. Proceedings of the National Academy of Sciences **96**:6529–6534.

Zhang, H. M., and B. G. Forde. 1998. An Arabidopsis MADS box gene that controls nutrient-induced changes in root architecture. Science **279**:407–409.

Zhang, H. X., and E. Blumwald. 2001. Transgenic salt-tolerant tomato plants accumulate salt in foliage but not in fruit. Nature Biotechnology **19**:765–768.

Zhang, Y., and J. G. Turner. 2008. Wound-induced endogenous jasmonates stunt plant growth by inhibiting mitosis. PLoS One **3**:e3699.

Zhang, Z. P., and I. T. Baldwin. 1997. Transport of [2-C-14]jasmonic acid from leaves to roots mimics wound-induced changes in endogenous jasmonic acid pools in *Nicotiana sylvestris*. Planta **203**:436–441.

Zhang, Z. Z., Y. B. Li, L. Qi, and X. C. Wan. 2006. Antifungal activities of major tea leaf volatile constituents toward *Colletorichum camelliae* Massea. Journal of Agricultural and Food Chemistry **54**:3936-3940.

Zipfel, C., and G. Felix. 2005. Plants and animals: a different taste for microbes? Current Opinion in Plant Biology **8**:353–360.

Zuluaga, D. L., S. Gonzali, E. Loreti, C. Pucciariello, E. Degl'Innoceni, L. Guidi, A. Alpi, and P. Perata. 2008. *Arabidopsis thaliana* MYB75/PAP1 transcription factor induces anthocyanin production in transgenic tomato plants. Functional Plant Biology **35**:606–618.

Index